# A FIRST COURSE IN
# FUNCTIONAL ANALYSIS

# A FIRST COURSE IN
# FUNCTIONAL ANALYSIS

### ORR MOSHE SHALIT

TECHNION - ISRAEL INSTITUTE OF TECHNOLOGY

HAIFA, ISRAEL

CRC Press
Taylor & Francis Group
Boca Raton  London  New York

CRC Press is an imprint of the
Taylor & Francis Group, an **informa** business
A CHAPMAN & HALL BOOK

CRC Press
Taylor & Francis Group
6000 Broken Sound Parkway NW, Suite 300
Boca Raton, FL 33487-2742

ISBN-13: 978-1-4987-7151-0 (hbk)
ISBN-13: 978-0-367-65813-7 (pbk)

---

**Library of Congress Cataloging-in-Publication Data**

---

Names: Shalit, Orr Moshe.
Title: A first course in functional analysis / Orr Moshe Shalit.
Description: Boca Raton : CRC Press, [2016] | Includes bibliographical references and
  index.
Identifiers: LCCN 2016045930| ISBN 9781498771610 (hardback : alk. paper) | ISBN
  9781315367132 (ebook) | ISBN 9781498771627 (ebook) | ISBN  9781498771641 (ebook)
  | ISBN 9781315319933 (ebook)
Subjects: LCSH: Functional analysis--Textbooks.
Classification: LCC QA320 .S45927 2016 | DDC 515/.7--dc23
LC record available at https://lccn.loc.gov/2016045930

---

**Visit the Taylor & Francis Web site at
http://www.taylorandfrancis.com**

**and the CRC Press Web site at
http://www.crcpress.com**

*To my mother and my father, Malka and Meir Shalit*

# Contents

# Preface

## In a nutshell

The purpose of this book is to serve as the accompanying text for a first course in functional analysis, taken typically by second- and third-year undergraduate students majoring in mathematics.

As I prepared for my first time teaching such a course, I found nothing among the countless excellent textbooks in functional analysis available that perfectly suited my needs. I ended up writing my own lecture notes, which evolved into this book (an earlier version appeared on my blog [31]).

The main goals of the course this book is designed to serve are to introduce the student to key notions in functional analysis (complete normed spaces, bounded operators, compact operators), alongside significant applications, with a special emphasis on the Hilbert space setting. The emphasis on Hilbert spaces allows for a rapid development of several topics: Fourier series and the Fourier transform, as well as the spectral theorem for compact normal operators on a Hilbert space. I did not try to give a comprehensive treatment of the subject, the opposite is true. I did my best to arrange the material in a coherent and effective way, leaving large portions of the theory for a later course. The students who finish this course will be ready (and hopefully, eager) for further study in functional analysis and operator theory, and will have at their disposal a set of tools and a state of mind that may come in handy in any mathematical endeavor they embark on.

The text is written for a reader who is either an undergraduate student, or the instructor in a particular kind of undergraduate course on functional analysis. The background required from the undergraduate student taking this course is minimal: basic linear algebra, calculus up to Riemann integration, and some acquaintance with topological and metric spaces (in fact, the basics of metric spaces will suffice; and all the required material in topology/metric spaces is collected in the appendix).

Some "mathematical maturity" is also assumed. This means that the readers are expected to be able to fill in some details here and there, not freak out when bumping into a slight abuse of notation, and so forth.

## More details on the contents and on some choices made

This book is tailor-made to accompany the course *Introduction to Functional Analysis* given at the Technion — Israel Institute of Technology. The official syllabus of the course is roughly: basic notions of Hilbert spaces and Banach spaces, bounded operators, Fourier series and the Fourier transform, the Stone-Weierstrass theorem, the spectral theorem for compact normal operators on a Hilbert space, and some applications. A key objective, not less important than the particular theorems taught, is to convey some underlying principles of modern analysis.

The design was influenced mainly by the official syllabus, but I also took into account the relative place of the course within the curriculum. The background that I could assume (mentioned above) did not include courses on Lebesgue integration or complex analysis. Another thing to keep in mind was that besides this course, there was no other course in the mathematics undergraduate curriculum giving a rigorous treatment of Fourier series or the Fourier transform. I therefore had to give these topics a respectable place in class. Finally, I also wanted to keep in mind that students who will continue on to graduate studies in analysis will take the department's graduate course on functional analysis, in which the Hahn-Banach theorems and the consequences of Baire's theorem are treated thoroughly. This allowed me to omit these classical topics with a clean conscience, and use my limited time for a deeper study in the context of Hilbert spaces (weak convergence, inverse mapping theorem, spectral theorem for compact normal operators), including some significant applications (PDEs, Hilbert function spaces, Pick interpolation, the mean ergodic theorem, integral equations, functional equations, Fourier series and the Fourier transform).

An experienced and alert reader might have recognized the inherent pitfall in the plan: how can one give a serious treatment of $L^2$ spaces, and in particular the theory of Fourier series and the Fourier transform, without using the Lebesgue integral? This is a problem which many instructors of introductory functional analysis face, and there are several solutions which can be adopted.

In some departments, the problem is eliminated altogether, either by making a course on Lebesgue integration a prerequisite to a course on functional analysis, or by keeping the introductory course on functional analysis free of $L^p$ spaces, with the main examples of Banach spaces being sequence spaces or spaces of continuous functions. I personally do not like either of these easy solutions. A more pragmatic solution is to use the Lebesgue integral as much as is needed, and to compensate for the students' background by either giving a crash course on Lebesgue integration or by waving one's hands where the going gets tough.

I chose a different approach: hit the problem head on using the tools available in basic functional analysis. I define the space $L^2[a, b]$ to be the completion of the space of piecewise continuous functions on $[a, b]$ equipped with the norm $\|f\|_2 = (\int_a^b |f(t)|^2 dt)^{1/2}$, which is defined in terms of the familiar Riemann integral. We can then use the Hilbert space framework to derive analytic results, such as convergence of Fourier series of elements in $L^2[a, b]$, and in particular we can get results on Fourier series for honest functions, such as $L^2$ convergence for piecewise continuous functions, or uniform convergence for periodic and $C^1$ functions.

Working in this fashion may seem clumsy when one is already used to working with the Lebesgue integral, but, for many applications to analysis it suffices. Moreover, it shows some of the advantages of taking a functional analytic point of view. I did not invent the approach of defining $L^p$ spaces as completions of certain space of nice functions, but I think that this book is unique in the extent to which the author really adheres to this approach: once the spaces are defined this way, we never look back, and *everything* is done with no measure theory.

To illustrate, in Section 8.2.2 we prove the mean ergodic theorem. A measure preserving composition operator on $L^2[0,1]$ is defined first on the dense subspace of continuous functions, and then extended by continuity to the completion. The mean ergodic theorem is proved by Hilbert space methods, as a nice application of some basic operator theory. The statement (see Theorem 8.2.5) in itself is significant and interesting even for piecewise continuous functions — one does not need to know the traditional definition of $L^2$ in order to appreciate it.

Needless to say, this approach was taken because of pedagogical constraints, and I encourage all my students to take a course on measure theory if they are serious about mathematics, *especially* if they are interested in functional analysis. The disadvantages of the approach we take to $L^2$ spaces are highlighted whenever we stare them in the face; for example, in Section 5.3, where we obtain the existence of weak solutions to PDEs in the plane, but fall short of showing that weak solutions are (in some cases) solutions in the classical sense.

The choice of topics and their order was also influenced by my personal teaching philosophy. For example, Hilbert spaces and operators on them are studied before Banach spaces and operators on them. The reasons for this are (a) I wanted to get to significant applications to analysis quickly, and (b) I do not think that there is a point in introducing greater generality before one can prove significant results in that generality. This is surely not the most efficient way to present the material, but there are plenty of other books giving elegant and efficient presentations, and I had no intention — nor any hope — of outdoing them.

## How to use this book

A realistic plan for teaching this course in the format given at the Technion (13 weeks, three hours of lectures and one hour of exercises every week) is to use the material in this book, in the order it appears, from Chapter 1 up to Chapter 12, skipping Chapters 6 and 11. In such a course, there is often time to include a section or two from Chapters 6 or 11, as additional illustrative applications of the theory. Going through the chapters in the order they appear, skipping chapters or sections that are marked by an asterisk, gives more or less the version of the course that I taught.

In an undergraduate program where there is a serious course on harmonic analysis, one may prefer to skip most of the parts on Fourier analysis (except $L^2$ convergence of Fourier series), and use the rest of the book as a basis for the course, either giving more time for the applications, or by teaching the material in Chapter 13 on the Hahn-Banach theorems. I view the chapter on the Hahn-Banach theorems as the first chapter in further studies in functional analysis. In the course that I taught, this topic was given as supplementary reading to highly motivated and capable students.

There are exercises spread throughout the text, which the students are expected to work out. These exercises play an integral part in the development of the material. Additional exercises appear at the end of every chapter. I recommend for the student, as well as the teacher, to read the additional exercises, because some of them contain interesting material that is good to know (e.g., Gibbs phenomenon, von Neumann's inequality, Hilbert-Schmidt operators). The teaching assistant will also find among the exercises some material better suited for tutorials (e.g., the solution of the heat equation, or the diagonalization of the Fourier transform). There is no solutions manual, but I invite any instructor who uses this book to teach a course, to contact me if there is an exercise that they cannot solve. With time I may gradually compile a collection of solutions to the most difficult problems.

Some of the questions are original, most of them are not. Having been a student and a teacher in functional and harmonic analysis for several years, I have already seen many similar problems appearing in many places, and some problems are so natural to ask that it does not seem appropriate to try to trace who deserves credit for "inventing" them. I only give reference to questions that I deliberately "borrowed" in the process of preparing this book. The same goes for the body of the material: most of it is standard, and I see no need to cite every mathematician involved; however, if a certain reference influenced my exposition, credit is given.

The appendix contains all the material from metric and topological spaces that is used in this book. Every once in while a serious student — typically majoring in physics or electrical engineering — comes and asks if he or she can take this course without having taken a course on metric spaces. The answer is: yes, if you work through the appendix, there should be no problem.

## Additional reading and alternative texts

There are countless good introductory texts on functional analysis and operator theory, and the bibliography contains a healthy sample. As a student and later as a teacher of functional analysis, I especially enjoyed and was influenced by the books by Gohberg and Goldberg [12], Devito [6], Kadison and Ringrose [16], Douglas [8], Riesz and Sz.-Nagy [26], Rudin [27], Arveson [3], Reed and Simon [24], and Lax [18]. These are all recommended, but only the first two are appropriate for a beginner. As a service to the reader, let me mention three more recent elementary introductions to functional analysis, by MacCluer [19], Hasse [13] and Eidelman, Milman and Tsolomitis [9]. Each one of these looks like an excellent choice for a textbook to accompany a first course.

I want to acknowledge that while working on the book I also made extensive use of the Web (mostly Wikipedia, but also MathOverflow/StackExchange) as a handy reference, to make sure I got things right, e.g., verify that I am using commonly accepted terminology, find optimal phrasing of a problem, etc.

## Acknowledgments

This book could not have been written without the support, encouragement and good advice of my beloved wife, Nohar. Together with Nohar, I feel exceptionally lucky and thankful for our dear children: Anna, Tama, Gev, Em, Shem, Asher and Sarah.

I owe thanks to many people for reading first drafts of these notes and giving me feedback. Among them are Alon Gonen, Shlomi Gover, Ameer Kassis, Amichai Lampert, Eliahu Levy, Daniel Markiewicz, Simeon Reich, Eli Shamovich, Yotam Shapira, and Baruch Solel. I am sorry that I do not remember the names of all the students who pointed out a mistake here or there, but I do wish to thank them all. Shlomi Gover and Guy Salomon also contributed a number of exercises. A special thank you goes to Michael Cwikel, Benjamin Passer, Daniel Reem and Guy Salomon, who have read large portions of the notes, found mistakes, and gave me numerous and detailed suggestions on how to improve the presentation.

I bet that after all the corrections by friends and students, there are still some errors here and there. Dear reader: if you find a mistake, please let me know about it! I will maintain a page on my personal website (currently http://oshalit.net.technion.ac.il) in which I will collect corrections.

I am grateful to Sarfraz Khan from CRC Press for contacting me and inviting me to write a book. I wish to thank Sarfraz, together with Michele Dimont the project editor, for being so helpful and kind throughout. I also owe many thanks to Samar Haddad the proofreader, whose meticulous work greatly improved the text.

My love for the subject and my point of view on it were strongly shaped by my teachers, and in particular by Boris Paneah (my Master's thesis advisor) and Baruch Solel (my Ph.D. thesis advisor). If this book is any good, then these men deserve much credit.

My parents, Malka and Meir Shalit, have raised me to be a man of books. This one, my first, is dedicated to them.

Haifa, 2017

## Notes for second printing, 2021

This revised printing incorporates corrections and improvements which have occurred to me after publication or have been suggested by various readers and friends. Special thanks go to Daniel Reem and Ross Pinsky, who provided extensive feedback. I also wish to thank Callum Fraser for his guidance in the production of the second printing.

Rosh Pina, 2021

# Chapter 1

## Introduction and the Stone-Weierstrass theorem

## 1.1 Background and motivation

Welcome to *A First Course in Functional Analysis*. What is functional analysis? A short answer is that functional analysis is a language and framework in which to formulate and study problems in analysis from an abstract point of view. In this introductory chapter, we will try to give a more detailed explanation of what is functional analysis. Along the way, we will prove the first significant theorem in this book: the Stone-Weierstrass theorem.

Functional analysis was born around 1900, when the mathematical climate was becoming suitable for an abstract and unified treatment of analytical objects. Consider, for example, the following *integral equation*:

$$f(x) + \int_a^b k(t, x) f(t) dt = g(x). \tag{1.1}$$

This equation was considered by I. Fredholm in 1903, and, like many important mathematical problems of the time, it arose from mathematical physics. In the equation above, the functions $g \in C_{\mathbb{R}}([a, b])$ and $k \in C_{\mathbb{R}}([a, b] \times [a, b])$ are given continuous functions, and $f$ is an unknown function (here and below $C_{\mathbb{R}}(X)$ denotes the space of continuous, real-valued functions on a topological space $X$). Fixing the function $k$, there are three basic questions one can ask about such equations:

1. **Solvability.** Does there exist a continuous solution $f$ to this equation given $g$? For what $g$ does a solution exist?

2. **Uniqueness.** Is the solution unique (when it exists)? Given a particular solution to the equation, can we describe the space of all solutions, or at least can we tell how "big" it is?

3. **Method of solution.** What is the solution? In other words, given $g$ can we write down a formula for the solution $f$, or at least describe a method of obtaining $f$ approximately?

Questions of a similar type are dealt with in a course in linear algebra,

1

when considering a system of linear equations $Ax = b$, where $x$ and $b$ are vectors in $\mathbb{R}^n$, and $A$ is an $n \times n$ matrix. There are many nontrivial things to say regarding the solvability of the equation $Ax = b$ which do not require knowing the specific matrix $A$, for example: if the equation $Ax = 0$ has a unique solution (namely, $x = 0$), then $Ax = b$ has a unique solution for *any* $b \in \mathbb{R}^n$. In the same vein, one is interested not only in answering the above three questions for the integral equation (1.1) given a particular $k$; it is also of interest to understand the unifying characteristics of equations of this type.

One may develop an ad hoc theory of integral equations, but it is most enlightening for us to put the above rather concrete equation in an abstract and general framework. Historically as well as methodologically, this is the starting point of functional analysis.

Let us try to force the above analytic problem into the framework of linear algebra. First, we note that $C_{\mathbb{R}}([a,b])$ is a real vector space. Next we notice, following Fredholm, that the above equation can be written as $(I + K)f = g$, where $I : C_{\mathbb{R}}([a,b]) \to C_{\mathbb{R}}([a,b])$ is the identity operator and $K : C_{\mathbb{R}}([a,b]) \to C_{\mathbb{R}}([a,b])$ is the so-called *integral operator*

$$Kf(x) = \int_a^b k(t,x)f(t)dt.$$

Finally, we observe that the operator $K$, and hence $I + K$, is a *linear* operator. Thus our problem is "just" that of solving a linear equation in the vector space $C_{\mathbb{R}}([a,b])$. However, these observations do not bring us much closer to being able to answer the above questions; they only suggest that it might be fruitful to study linear operators on infinite dimensional vector spaces.

Linear operators on infinite dimensional spaces turn out to be too large a class to be treated in a meaningful manner. We will concentrate on the study of continuous or compact operators acting on complete normed vector spaces. In this setting significant results can be obtained; for example, in Chapter 9 we will see that the equation $(I + K)f = g$ has a unique solution for any $g$, if and only if the equation $(I + K)f = 0$ has a unique solution. Methods of finding the solution will also be developed.

Note that, unlike in the case of a linear system of equations $Ax = b$, the equation (1.1) makes sense within many different vector spaces. Would it be easier to solve the problem if we assumed that all functions are differentiable? The equation makes sense for integrable $f$ — maybe we should consider $I + K$ as a linear operator on the larger space of all integrable functions, or maybe on the space of square integrable functions? Do the basic properties of the operator $I + K$ change if we think of it as an operator on the bigger space? We will see that considering the integral equation as an operator equation in the space of *square integrable functions* on $[a,b]$ does not change some characteristic features of the problem, while on the other hand it facilitates actually solving the problem.

The space of square integrable functions on an interval is an example of a *Hilbert space*. Hilbert spaces are the infinite dimensional spaces that are closest to finite dimensional spaces, and they are the most tractable. What makes them so tractable are the fact that they have an inner product, and the fact that they are complete metric spaces. In this book, a special emphasis is put on Hilbert spaces, and in this setting integral equations are best understood. We will begin our study of Hilbert spaces in the next chapter.

The goal of this book is to present a set of tools, ideas and results that can be used to understand linear operators on infinite dimensional spaces. But before we can develop a theory of linear operators on infinite dimensional spaces, we must study the spaces themselves. We will have to wait until much later in the book, before we can prove significant results on the solvability of integral equations; in Chapter 11 we will complete our treatment of this subject. For now, we part from equation (1.1), and we take a closer look at the space $C_{\mathbb{R}}([a, b])$.

---

## 1.2   The Weierstrass approximation theorem

In linear algebra one learns that every finite dimensional vector space has a basis. This fact is incredibly useful, both conceptually and also from a practical point of view; just consider how easy it is for us to think about $\mathbb{R}^n$ having at hand the standard basis $e_1 = (1, 0, \ldots, 0), \ldots, e_n = (0, \ldots, 0, 1)$. In particular, the existence of bases is crucial to our understanding of linear maps and of the solvability of linear equations of the form $Ax = b$.

In analysis we encounter many vector spaces that are infinite dimensional. For example, the space $C_{\mathbb{R}}([a, b])$ is an infinite dimensional real vector space. Our experience in linear algebra suggests that it would be useful to have some kind of basis for studying this space, and in particular it should be helpful for considering linear equations as above.

A standard application of Zorn's lemma shows that every vector space $V$ has a basis — a set $\{u_i\}_{i \in I} \subset V$ such that for all nonzero $v \in V$, there is a unique choice of finitely many distinct indices $i_1, \ldots, i_k \in I$ and nonzero scalars $c_1, \ldots, c_k$ satisfying

$$v = c_1 u_{i_1} + \ldots + c_k u_{i_k}.$$

In the context of analysis, such a basis is called a **Hamel basis**, to distinguish it from other notions of basis which are used. For most infinite dimensional vector spaces a Hamel basis is almost useless, because of the following fact.

**Exercise 1.2.1.** A Hamel basis for $C_{\mathbb{R}}([a, b])$ must be uncountable (for a hint see Exercise 7.5.16).

The fact that Hamel bases for $C_{\mathbb{R}}([a, b])$ must be uncountable suggests that we may want to relax our notion of basis. By the exercise, we have no hope to find a countable basis. However, we can hope for the next best thing, which is a linearly independent sequence of functions $\{f_n\}_{n=0}^{\infty} \subset C_{\mathbb{R}}([a,b])$ which in some sense span the space. A natural guess would be that the sequence of monomials $f_n(t) = t^n$ are as close to being a basis for $C_{\mathbb{R}}([a,b])$ as one can reasonably expect. This is true, as Weierstrass proved in 1885.

**Theorem 1.2.2** (Weierstrass's approximation theorem). *Let $f : [a,b] \to \mathbb{R}$ be a continuous function. For every $\epsilon > 0$, there exists a polynomial $p$ with real coefficients, such that for all $t \in [a,b]$,*

$$|p(t) - f(t)| < \epsilon.$$

Since not every continuous function is a polynomial, we cannot hope to obtain every continuous function as a linear combination of monomials. Weierstrass's approximation theorem assures us that we can at least use linear combinations of monomials to approximate every continuous function to any given precision. Thus the monomials can be said to "generate" $C_{\mathbb{R}}([a,b])$.

Are there any other sequences of nice functions that generate $C_{\mathbb{R}}([a,b])$? To give another example, we need a couple of definitions. A $\mathbb{Z}$-*periodic function* is a function $f : \mathbb{R} \to \mathbb{R}$ such that $f(x+n) = f(x)$ for all $x \in \mathbb{R}$ and all $n \in \mathbb{Z}$. A *trigonometric polynomial* is a function $q$ of the form

$$q(x) = \sum_{n=0}^{N} a_n \cos(2\pi n x) + b_n \sin(2\pi n x).$$

Later in this book, when we will study Fourier series, we will require the following theorem (see also Exercise 1.6.7). Like the previous theorem, this one is also due to Weierstrass.

**Theorem 1.2.3** (Trigonometric approximation theorem). *Let $f : \mathbb{R} \to \mathbb{R}$ be a continuous $\mathbb{Z}$-periodic function. For every $\epsilon > 0$, there exists a trigonometric polynomial $q$, such that for all $t \in \mathbb{R}$,*

$$|q(t) - f(t)| < \epsilon.$$

It turns out that the most elegant way to obtain the above two theorems is to consider a more general problem. We will obtain both Theorem 1.2.2 and Theorem 1.2.3 as simple consequences of the Stone-Weierstrass theorem, which is a broad generalization of these theorems. Our goal in the remainder of this chapter is to present and prove the Stone-Weierstrass theorem. Besides obtaining useful results for later purposes in the course, this also serves to highlight the spirit of functional analysis. (We will return to the problem of finding a good notion of basis for infinite dimensional spaces in Chapter 3.)

## 1.3    The Stone-Weierstrass theorem

Let $X$ be a compact Hausdorff topological space (the appendix contains all the material in topological and metric spaces that is required for this book). We will let $C_{\mathbb{R}}(X)$ denote the space of continuous, real-valued functions on $X$; likewise, $C(X)$ denotes the space of continuous, complex-valued functions on $X$. On both of these spaces we define the **supremum norm** of a function $f$ to be

$$\|f\|_\infty = \sup_{x \in X} |f(x)|.$$

The quantity $d(f,g) = \|f - g\|_\infty$ defines a metric on $C_{\mathbb{R}}(X)$ (and also on $C(X)$) and is considered to be the distance between the two functions $f$ and $g$. This distance makes both $C_{\mathbb{R}}(X)$ and $C(X)$ into complete metric spaces. Both these spaces are vector spaces over the appropriate field, with the usual operations of pointwise addition of functions and scalar multiplication. In fact, if $f, g \in C_{\mathbb{R}}(X)$, then the pointwise product $fg$ is also in $C_{\mathbb{R}}(X)$, and together with the vector space operations, this gives $C_{\mathbb{R}}(X)$ the structure of an **algebra**.

**Definition 1.3.1.** If $A$ is a subspace of $C_{\mathbb{R}}(X)$ or of $C(X)$, then it is said to be a **subalgebra** if for all $f, g \in A$, $fg$ is also in $A$.

**Definition 1.3.2.** A subalgebra $A \subseteq C_{\mathbb{R}}(X)$ is said to **separate points** if for every pair of distinct points $x, y \in X$ there exists some $f \in A$ such that $f(x) \neq f(y)$.

**Theorem 1.3.3** (Stone-Weierstrass theorem (real version)). *Let $A$ be a closed subalgebra of $C_{\mathbb{R}}(X)$ which contains the constant functions and separates points. Then $A = C_{\mathbb{R}}(X)$.*

One obtains Theorem 1.2.2 immediately by letting $X = [a, b]$ and taking $A$ to be the closure of the algebra of polynomials with respect to the supremum norm, noting that the norm closure of an algebra is an algebra.

**Exercise 1.3.4.** Let $A \subseteq C_{\mathbb{R}}(X)$ be a subalgebra, and let $\overline{A}$ be its closure. Then $\overline{A}$ is also a subalgebra.

Another convenient way of stating the Stone-Weierstrass theorem is given in the following exercise.

**Exercise 1.3.5.** Prove that Theorem 1.3.3 is equivalent to the following statement: *If $A$ is a subalgebra of $C_{\mathbb{R}}(X)$ which contains the constant functions and separates points, then $\overline{A} = C_{\mathbb{R}}(X)$.*

To obtain the trigonometric approximation theorem (Theorem 1.2.3), one first notes that, due to standard trigonometric identities, the trigonometric polynomials form an algebra. Next, one needs to realize that the continuous, real-valued, $\mathbb{Z}$-periodic functions on $\mathbb{R}$ can be identified with the continuous, real-valued functions on the torus $\mathbb{T} = \{z \in \mathbb{C} \mid |z| = 1\}$. Indeed, the mapping $\Phi : C_{\mathbb{R}}(\mathbb{T}) \to C_{\mathbb{R}}(\mathbb{R})$ given by

$$\Phi(f)(t) = f\left(e^{2\pi i t}\right) \quad , \quad t \in \mathbb{R}$$

maps $C_{\mathbb{R}}(\mathbb{T})$ onto the algebra $C_{\mathbb{R},per}(\mathbb{R})$ of continuous $\mathbb{Z}$-periodic functions. Moreover, $\Phi$ maps the constant function to the constant function, it is linear, and it respects multiplication: $\Phi(fg) = \Phi(f)\Phi(g)$. Therefore, the same holds for the inverse $\Phi^{-1} : C_{\mathbb{R},per}(\mathbb{R}) \to C_{\mathbb{R}}(\mathbb{T})$. Finally, $\sup_{t \in \mathbb{R}} |\Phi(f)(t)| = \sup_{|z|=1} |f(z)|$ for all $f \in C_{\mathbb{R}}(\mathbb{T})$.

Now put $X = \mathbb{T}$, and take $A \subseteq C_{\mathbb{R}}(\mathbb{T})$ to be the inverse image of all trigonometric polynomials under $\Phi$ (if we identify $C_{\mathbb{R}}(\mathbb{T})$ and $C_{\mathbb{R},per}(\mathbb{R})$, then under this identification $A$ is simply the algebra of all trigonometric polynomials, but we have chosen to make this identification explicit with the use of the map $\Phi$). By the previous paragraph, in order to prove Theorem 1.2.3 it suffices to show that $A$ is dense in $C_{\mathbb{R}}(X)$. It is elementary to check that $A$ contains the constants and separates points on $X = \mathbb{T}$. Applying the version of the Stone-Weierstrass theorem given in Exercise 1.3.5, we find that $A$ is dense in $C_{\mathbb{R}}(\mathbb{T})$, therefore $\Phi(A)$ is dense in $C_{\mathbb{R},per}(\mathbb{R})$. That concludes the proof of the trigonometric approximation theorem.

**Exercise 1.3.6.** Fill in the details in the proof of Theorem 1.2.3. In particular, prove that $A$ is an algebra that separates points, and prove that $\Phi$ is surjective.

### Proof of the Stone-Weierstrass theorem

Let $A \subseteq C_{\mathbb{R}}(X)$ be as in the statement of Theorem 1.3.3. We isolate a few lemmas before reaching the main argument of the proof.

**Lemma 1.3.7.** *On every interval $[-L, L]$, the absolute value function is uniformly approximable by polynomials. That is, for every $\epsilon > 0$, there exists a polynomial $p$ such that*

$$|p(t) - |t|| < \epsilon \tag{1.2}$$

*for all $t \in [-L, L]$.*

*Proof.* It suffices to prove the lemma for the interval $[-1, 1]$ (why?). To this end, consider the function $h(x) = (1-x)^{1/2}$. It is a standard (but nontrivial) exercise in first-year analysis to show that the Taylor series of $h$ about the point $0$ converges uniformly in the closed interval $[-1, 1]$. Truncating the series at some high power, we find, given $\epsilon > 0$, a polynomial $q$ such that $|q(x) - h(x)| < \epsilon$ for all $x \in [-1, 1]$. Now since $|t| = h(1 - t^2)$, the polynomial $p(t) = q(1 - t^2)$ satisfies (1.2). $\qquad \square$

**Exercise 1.3.8.** Fill in the details of the above proof; in particular prove the uniform convergence of the Maclaurin series of $h$ on $[-1, 1]$. (**Hint:** use the integral form of the remainder for the Taylor polynomial approximation.)

For every function $f$, we let $|f|$ denote the function $|f| : x \mapsto |f(x)|$.

**Lemma 1.3.9.** *If $f \in A$, then the function $|f|$ is also in $A$.*

*Proof.* Let $\epsilon > 0$ be given. We will find a function $g \in A$ such that $\|g - |f|\|_\infty < \epsilon$. Since $A$ is closed and since $\epsilon$ is arbitrary, this will show that $|f| \in A$.

Let $I = [-\|f\|_\infty, \|f\|_\infty]$. By the previous lemma there exists a polynomial $p$ such that $\sup_{t \in I} |p(t) - |t|| < \epsilon$. Put $g = p \circ f$. Since $A$ is an algebra and $p$ is a polynomial, $g \in A$. Thus

$$\|g - |f|\|_\infty = \sup_{x \in X} |p(f(x)) - |f(x)|| \leq \sup_{t \in I} |p(t) - |t|| < \epsilon,$$

as required. □

For any two functions $f, g$, we let $f \wedge g$ and $f \vee g$ denote the functions $f \wedge g : x \mapsto \min\{f(x), g(x)\}$ and $f \vee g : x \mapsto \max\{f(x), g(x)\}$.

**Lemma 1.3.10.** *If $f, g \in A$, then the functions $f \wedge g$ and $f \vee g$ are also in $A$.*

*Proof.* This follows immediately from Lemma 1.3.9 together with the formulas $\min\{a, b\} = \frac{a+b-|a-b|}{2}$ and $\max\{a, b\} = \frac{a+b+|a-b|}{2}$, which hold true for all real $a$ and $b$. □

**Lemma 1.3.11.** *For every pair of distinct points $x, y \in X$, and every $a, b \in \mathbb{R}$, there exists a function $g \in A$ such that $g(x) = a$ and $g(y) = b$.*

*Proof.* Exercise. □

**Completion of the proof of the Stone-Weierstrass theorem.** Let $f \in C_\mathbb{R}(X)$. We must show that $f \in A$. It suffices, for a fixed $\epsilon > 0$, to find $h \in A$ such that $\|f - h\|_\infty < \epsilon$.

We start by choosing, for every $x, y \in X$, a function $f_{xy} \in A$ such that $f_{xy}(x) = f(x)$ and $f_{xy}(y) = f(y)$. This is possible thanks to Lemma 1.3.11.

Next we produce, for every $x \in X$, a function $g_x \in A$ such that $g_x(x) = f(x)$ and $g_x(y) < f(y) + \epsilon$ for all $y \in X$. This is done as follows. For every $y \in X$, let $U_y$ be an open neighborhood of $y$ in which $f_{xy} < f + \epsilon$. The compactness of $X$ ensures that there are finitely many of these neighborhoods, say $U_{y_1}, \ldots, U_{y_m}$, that cover $X$. Then $g_x = f_{xy_1} \wedge \ldots \wedge f_{xy_m}$ does the job ($g_x$ is in $A$, thanks to Lemma 1.3.10).

Finally, we find $h \in A$ such that $|h(x) - f(x)| < \epsilon$ for all $x \in X$. For every $x \in X$ let $V_x$ be an open neighborhood of $x$ where $g_x > f - \epsilon$. Again we find a finite cover $V_{x_1}, \ldots, V_{x_n}$ and then define $h = g_{x_1} \vee \ldots \vee g_{x_n}$. This function lies between $f + \epsilon$ and $f - \epsilon$, so it satisfies $|h(x) - f(x)| < \epsilon$ for all $x \in X$, and the proof is complete. □

**Exercise 1.3.12.** Did we use the assumption that $X$ is Hausdorff? Explain.

## 1.4    The Stone-Weierstrass theorem over the complex numbers

Often, one finds it more convenient to study or to use the algebra $C(X)$ of continuous complex-valued functions on $X$. Maybe the reader has not encountered this algebra of functions before, but its structure as a vector space is very close to that of $C_{\mathbb{R}}(X)$.

**Exercise 1.4.1.** Prove that $C(X) = \{u + iv \mid u, v \in C_{\mathbb{R}}(X)\}$.

If $f = u + iv$ where $u, v \in C_{\mathbb{R}}(X)$, then we denote $\operatorname{Re}f = u$ and $\operatorname{Im}f = v$. Thus, for every $f \in C(X)$, $f = \operatorname{Re}f + i\operatorname{Im}f$.

It turns out that it is harder for a subalgebra of $C(X)$ to be dense in $C(X)$ than it is for a subalgebra of $C_{\mathbb{R}}(X)$ to be dense in $C_{\mathbb{R}}(X)$. Consider the following example.

**Example 1.4.2.** Let $\mathbb{D}$ denote the open unit disc in $\mathbb{C}$, and let $\overline{\mathbb{D}}$ denote its closure. Let $A(\mathbb{D})$ denote the **disc algebra**, which is defined to be the closure of complex polynomials in $C(\overline{\mathbb{D}})$, that is

$$A(\mathbb{D}) = \overline{\left\{ z \mapsto \sum_{k=0}^{N} a_k z^k : N \in \mathbb{N}, a_0, \ldots, a_N \in \mathbb{C} \right\}}^{\|\cdot\|_\infty}.$$

Certainly, $A(\mathbb{D})$ is a complex algebra which contains the constants and separates points. By a theorem in complex analysis, the uniform limit of analytic functions is analytic. Thus, every element of $A(\mathbb{D})$ is analytic in $\mathbb{D}$, so this algebra is quite far from being the entire algebra $C(\overline{\mathbb{D}})$.

It is worth stressing that in the above example we mean polynomials in one complex variable $z$. We *do not mean* polynomials in the two variables $x$ and $y$, where $z = x + iy = \operatorname{Re}z + i\operatorname{Im}z$.

**Exercise 1.4.3.** Is the space $\mathbb{C}[x, y]$ of two variable complex polynomials dense in $C(\overline{\mathbb{D}})$?

To make the Stone-Weierstrass theorem work in the complex-valued case, one needs to add one additional assumption.

**Definition 1.4.4.** A subspace $S \subseteq C(X)$ is said to be **self-adjoint** if for every $f \in S$, the complex conjugate of $f$ (i.e., the function $\overline{f} : x \mapsto \overline{f(x)}$) is also in $S$.

**Theorem 1.4.5** (Stone-Weierstrass theorem (complex version)). *Let $A$ be a closed and self-adjoint subalgebra of $C(X)$ which contains the constant functions and separates points. Then $A = C(X)$.*

*Proof.* Consider the real vector space $\mathrm{Re}A = \{\mathrm{Re}f : f \in A\}$. Since $\mathrm{Re}f = \frac{f+\bar{f}}{2}$ and $A$ is self-adjoint, it follows that $\mathrm{Re}A \subseteq A$. Because $A$ is a subalgebra of $C(X)$, $\mathrm{Re}A$ is a subalgebra of $C_{\mathbb{R}}(X)$. From closedness of $A$ it follows that $\mathrm{Re}A$ is closed, too.

From the assumption that $A$ is a subspace that separates points, it follows that $\mathrm{Re}A$ also separates points. Indeed, given $x, y \in X$, let $f \in A$ such that $f(x) \neq f(y)$. Then, either $\mathrm{Re}f(x) \neq \mathrm{Re}f(y)$, or $\mathrm{Im}f(x) \neq \mathrm{Im}f(y)$. But $\mathrm{Im}f = \mathrm{Re}(-if) \in \mathrm{Re}A$, so $\mathrm{Re}A$ separates points.

Thus, $\mathrm{Re}A$ is a closed, real subalgebra of $C_{\mathbb{R}}(X)$ that contains the constants and separates points. By the (real) Stone-Weierstrass theorem, $\mathrm{Re}A = C_{\mathbb{R}}(X)$. It follows that every real-valued continuous function on $X$ is in $A$. Symmetrically, every imaginary valued continuous function on $X$ is in $A$. By Exercise 1.4.1 we conclude that $C(X) = A$. $\qquad\square$

---

## 1.5   Concluding remarks

Functional analysis originated from an interest in solving analytical problems such as the integral equation (1.1). Equations of this kind can be rephrased as problems about operators acting on infinite dimensional vector spaces. If one wishes to understand operators on infinite dimensional vector spaces, the first thing to do is to study the spaces themselves. In this chapter we took a look at the space $C_{\mathbb{R}}([a,b])$, and proved Weierstrass's approximation theorem, which was a byproduct of the Stone-Weierstrass theorem. Besides obtaining important theorems to be used subsequently, these results should give a flavor of how life in infinite dimensional vector spaces is different from what one is used to in finite dimensional spaces.

The Stone-Weierstrass theorem and the way that we have applied it serve as an example of functional analysis at work. We had a concrete approximation problem — approximating continuous functions by polynomials or by trigonometric polynomials — which was solved by considering a vastly more general approximation problem. Considering a more general problem serves two purposes. First, after we have proved the result, we have a ready-to-use tool that will be applicable in many situations. Second, by generalizing the problem we strip away the irrelevant details (for example, the particular nature of the functions we are trying to approximate with or the nature of the space on which they live) and we are left with the essence of the problem.

To prove the theorem it was convenient to employ the language of abstract analysis, namely, to introduce a norm and to consider the problem inside an algebra which is also a metric space. It was convenient to consider a closed subalgebra $A$, even though there was no closed subalgebra in the original problems, and even though this closed subalgebra turned out to be the whole space of continuous functions.

In mathematical culture, people sometimes make a distinction between "hard analysis" and "soft analysis". "Hard analysis" usually refers to arguments that require explicit estimates or calculations (such as are required to show that a power series converges in an interval, or that some generalized integral converges conditionally), and "soft analysis" refers to arguments that use algebraic constructs or deep topological considerations. Because functional analysis provides an algebraic framework for a unified and elegant treatment of problems in analysis, many students naively consider functional analysis to be a magical device which turns every hard analytical problem into a soft one. Things are not that simple.

The elegant and abstract proof of the Weierstrass theorem presented above is quite softer than Weierstrass's original proof, which involved integration on the real line and the use of power series. However, we could not do without some piece of "hard analysis": we required the fact that the Taylor series of $h(x) = \sqrt{1-x}$ converges uniformly in $[-1,1]$. This is an instance of the following maxim (to be taken with a grain of salt): *there is no analysis without "hard analysis"*.

What is meant by this is that one will hardly ever get an interesting theorem of substance, which applies to concrete cases in analysis, that does not involve some kind of difficult calculation, delicate estimation, or at least a clever trick. In our example, the hard analysis part is showing that the power series of $\sqrt{1-x}$ converges uniformly in $[-1,1]$. The Stone-Weierstrass theorem, which has a very soft proof, then spares us the use of hard techniques when it gives us the polynomial and the trigonometric approximation theorems for free. But the hard analysis cannot be eliminated entirely. It is good to keep this in mind.

## 1.6   Additional exercises

**Exercise 1.6.1.** Let $f \in C_{\mathbb{R}}([-1,1])$ such that $f(0) = 0$. Prove that $f$ can be approximated in uniform norm by polynomials with zero constant coefficient.

**Exercise 1.6.2.** Let $C_0(\mathbb{R})$ be the algebra of all continuous functions that vanish at infinity, that is

$$C_0(\mathbb{R}) = \{f \in C(\mathbb{R}) : \lim_{x \to \infty} f(x) = \lim_{x \to -\infty} f(x) = 0\}.$$

Let $A$ be a self-adjoint subalgebra of $C_0(\mathbb{R})$ that separates points, and assume that for all $x \in \mathbb{R}$, there exists $f \in A$ such that $f(x) \neq 0$. Prove that $\overline{A} = C_0(\mathbb{R})$.

**Exercise 1.6.3.** Determine which of the following statements are true.

1. For every $f \in C_{\mathbb{R}}([a,b])$, all $x_1, x_2, \ldots, x_n \in [a,b]$, and every $\epsilon > 0$, there exists a polynomial $p$ such that $\sup_{x \in [a,b]} |p(x) - f(x)| < \epsilon$ and $p(x_i) = f(x_i)$ for $i = 1, 2, \ldots, n$.

2. For every $f \in C_{\mathbb{R}}([a,b])$ which has $n$ continuous derivatives on $[a,b]$, and every $\epsilon > 0$, there exists a polynomial $p$ such that $\sup_{x \in [a,b]} |p^{(k)}(x) - f^{(k)}(x)| < \epsilon$ for all $k = 0, 1, \ldots, n$. (Here $f^{(k)}$ denotes the $k$th derivative of $f$.)

What happens if we replace the word "polynomial" with "trigonometric polynomial"? You might need to reformulate the question so that it makes clearer sense.

**Exercise 1.6.4.** 1. Let $f$ be a continuous real-valued function on $[a,b]$ such that $\int_a^b f(x)x^n dx = 0$ for all $n = 0, 1, 2, \ldots$. Prove that $f \equiv 0$.

2. Let $f$ be a continuous real-valued function on $[-1, 1]$ such that $\int_{-1}^1 f(x)x^n dx = 0$ for all odd $n$. What can you say about $f$?

**Exercise 1.6.5.** Let $X$ be a compact metric space. Prove that $C_{\mathbb{R}}(X)$ is a separable metric space.

**Exercise 1.6.6.** Let $X$ and $Y$ be compact metric spaces. Let $X \times Y$ be the product space (see Definition A.3.10) and fix a function $h \in C(X \times Y)$. Prove that for every $\epsilon > 0$, there exist $f_1, \ldots, f_n \in C(X)$ and $g_1, \ldots, g_n \in C(Y)$ such that for all $(x, y) \in X \times Y$,

$$|h(x,y) - \sum_{i=1}^n f_i(x)g_i(y)| < \epsilon.$$

**Exercise 1.6.7.** A $\mathbb{Z}^k$-*periodic function* is a function $f : \mathbb{R}^k \to \mathbb{R}$ such that $f(x + n) = f(x)$ for all $x \in \mathbb{R}^k$ and $n \in \mathbb{Z}^k$. A *trigonometric polynomial* on $\mathbb{R}^k$ is a function $q$ of the form

$$q(x) = \sum_n a_n \cos(2\pi n \cdot x) + b_n \sin(2\pi n \cdot x),$$

where the sum above is finite, and for $x = (x_1, \ldots, x_k)$ and $n = (n_1, \ldots, n_k)$ we write $n \cdot x = \sum n_i x_i$. Prove the following theorem.

**Theorem** (Trigonometric approximation theorem). *Let $f : \mathbb{R}^k \to \mathbb{R}$ be a continuous $\mathbb{Z}^k$-periodic function. For every $\epsilon > 0$ there exists a trigonometric polynomial $q$ such that for all $x \in \mathbb{R}^k$,*

$$|q(x) - f(x)| < \epsilon.$$

**Exercise 1.6.8** (Holladay[1]). Let $X$ be a compact Hausdorff space and let $C_{\mathbb{H}}(X)$ denote the algebra of continuous functions taking quaternionic values. Here, $\mathbb{H}$ denotes the quaternions, that is, the four-dimensional real vector space

$$\mathbb{H} = \{a + bi + cj + dk : a, b, c, d \in \mathbb{R}\},$$

with a basis consisting of $1 \in \mathbb{R}$ and three additional elements $i, j, k$, and with multiplication determined by the multiplication of the reals and by the identities

$$i^2 = j^2 = k^2 = ijk = -1.$$

(The topology on $\mathbb{H}$ — required for the notion of continuous functions from $X$ into $\mathbb{H}$ — is the one given by identifying $\mathbb{H}$ with $\mathbb{R}^4$.) Prove that if $A$ is a closed subalgebra of $C_{\mathbb{H}}(X)$ that separates points and contains all the constant $\mathbb{H}$-valued functions, then $A = C_{\mathbb{H}}(X)$.

---

[1] This exercise was suggested to me by Shlomi Gover, who found the note [14] when he was a teaching assistant in the course I taught.

# Chapter 2

## Hilbert spaces

### 2.1  Background and motivation

In the previous chapter, we briefly discussed the problem of finding a useful basis for an infinite dimensional vector space. We saw how the sequence of monomials $\{t^n\}_{n=0}^{\infty}$ is in some sense a spanning set for $C_{\mathbb{R}}([a,b])$. The monomials are also linearly independent, so it may seem that this sequence deserves to be called a basis. However, it turns out that in infinite dimensional spaces the very notion of a basis requires much care. For instance we know that polynomials can be used to approximate any continuous function, but the approximating polynomial is not unique.

Recall that we introduced the metric $d(f,g) = \|f - g\|_{\infty}$ on $C_{\mathbb{R}}([a,b])$, and this helped us clarify in what sense $\{t^n\}_{n=0}^{\infty}$ can serve as a spanning set for $C_{\mathbb{R}}([a,b])$. But there are other distance functions which one can impose on $C_{\mathbb{R}}([a,b])$; for instance,

$$d_1(f,g) = \int_a^b |f(t) - g(t)| dt,$$

or

$$d_2(f,g) = \sqrt{\int_a^b |f(t) - g(t)|^2 dt}.$$

Is the supremum metric the best metric for studying the space $C_{\mathbb{R}}([a,b])$? For example, taking the integral equation

$$f(x) + \int_a^b k(t,x) f(t) dt = g(x)$$

as our starting point, it is not clear which metric will help us better at understanding the solvability of the problem. It turns out that for the purpose of understanding integral equations, it is profitable to consider the space $C_{\mathbb{R}}([a,b])$ as a metric space with the metric $d_2$. What makes $d_2$ useful is the fact that it is induced from an inner product.

At this point we abandon these vague and very general questions, and focus our attention on spaces in which the metric comes from an inner product. In this and in the next few chapters, we will study the basic theory of **Hilbert** *spaces*. Hilbert spaces are infinite dimensional inner product spaces that are also complete as metric spaces. The reason that we begin with Hilbert spaces is that familiar procedures from linear algebra can be easily generalized to this setting. In particular, we will see that Hilbert spaces admit orthonormal bases, and this fact will provide us very quickly with an application of the abstract theory to concrete problems in analysis.

---

## 2.2   The basic definitions

### 2.2.1   Inner product spaces

**Definition 2.2.1.** Let $G$ be a complex vector space. $G$ is said to be an ***inner product space*** if there exists a function $\langle \cdot, \cdot \rangle : G \times G \to \mathbb{C}$ that satisfies the following conditions:

1. $\langle f, f \rangle \geq 0$ for all $f \in G$.

2. $\langle f, f \rangle = 0$ if and only if $f = 0$.

3. $\langle f, g \rangle = \overline{\langle g, f \rangle}$ for all $f, g \in G$.

4. $\langle af + bg, h \rangle = a\langle f, h \rangle + b\langle g, h \rangle$ for all $f, g, h \in G$ and all $a, b \in \mathbb{C}$.

The function $\langle \cdot, \cdot \rangle : G \times G \to \mathbb{C}$ is referred to as an ***inner product***.

From properties 3 and 4 above, we have the following property too:

$$\langle h, af + bg \rangle = \overline{a}\langle h, f \rangle + \overline{b}\langle h, g \rangle \text{ for all } f, g, h \in G \text{ and all } a, b \in \mathbb{C}.$$

**Remark 2.2.2.** The vector spaces arising in functional analysis are usually over either the real numbers $\mathbb{R}$ or the complex numbers $\mathbb{C}$. In this text, we shall consider mostly complex spaces, that is, spaces over $\mathbb{C}$; this will keep statements clean. Almost all of the results discussed in this book hold equally well for real Hilbert spaces with similar proofs. We will try to take note of instances where the fact that we are working over $\mathbb{C}$ makes a difference.

Henceforth $G$ will denote an inner product space. Sometimes we will denote the inner product $\langle f, g \rangle_G$ for emphasis.

**Example 2.2.3.** The space $G = \mathbb{C}^n$ with the standard inner product:

$$\langle x, y \rangle = \sum_{i=1}^{n} x_i \overline{y}_i.$$

It is worth noting that as a metric space, $\mathbb{C}^n$ can be identified with $\mathbb{R}^{2n}$ in the obvious way.

**Example 2.2.4.** The space $G = C([0,1])$ (the continuous complex-valued functions on the interval) with the inner product

$$\langle f,g \rangle = \int_0^1 f(t)\overline{g(t)}dt.$$

The reader might not be familiar with Riemann integration of complex-valued functions. There is really nothing to it: if $f(t) = u(t) + iv(t)$ is a continuous, complex-valued function where $u$ and $v$ are its real and imaginary parts (defined by $u(t) = \operatorname{Re} f(t)$, $v(t) = \operatorname{Im} f(t)$), then

$$\int_a^b f(t)dt = \int_a^b u(t)dt + i\int_a^b v(t)dt.$$

It is straightforward to show that one obtains the same value for the integral if one takes the limit of Riemann sums.

**Definition 2.2.5.** If $f \in G$, one defines the **norm** of $f$ to be the nonnegative real number $\|f\|$ given by

$$\|f\| = \sqrt{\langle f,f \rangle}.$$

**Definition 2.2.6.** A **unit vector** in an inner product space is an element $f \in G$ such that $\|f\| = 1$.

The norm is defined in terms of the inner product, but one may also recover the inner product from the norm.

**Exercise 2.2.7.** Prove the **polarization identities**:

1. If $G$ is a real inner product space, then for all $f, g \in G$,

$$\langle f,g \rangle = \frac{1}{4}\left(\|f+g\|^2 - \|f-g\|^2\right).$$

2. If $G$ is a complex inner product space, then for all $f, g \in G$,

$$\langle f,g \rangle = \frac{1}{4}\left(\|f+g\|^2 - \|f-g\|^2 + i\|f+ig\|^2 - i\|f-ig\|^2\right).$$

**Theorem 2.2.8** (Cauchy-Schwarz inequality). *For all $f, g \in G$,*

$$|\langle f,g \rangle| \leq \|f\|\|g\|.$$

*Equality holds if and only if $f$ and $g$ are linearly dependent.*

*Proof.* If $f$ and $g$ are linearly dependent, then without loss of generality $f = cg$, and $|\langle f,g \rangle| = |c|\|g\|^2 = \|f\|\|g\|$.

If $f, g \in G$ are not linearly dependent (so in particular nonzero), then consider the unit vectors $u = \frac{1}{\|f\|}f$ and $v = \frac{1}{\|g\|}g$. We need to prove that

$|\langle u, v \rangle| < 1$. Let $t \in \mathbb{C}$ be such that $|t| = 1$ and $\bar{t}\langle u, v \rangle = -|\langle u, v \rangle|$. Using the properties of inner product, we have

$$
\begin{aligned}
0 < \|u + tv\|^2 &= \langle u + tv, u + tv \rangle \\
&= \langle u, u \rangle + \bar{t}\langle u, v \rangle + t\langle v, u \rangle + |t|^2\langle v, v \rangle \\
&= 2 - 2|\langle u, v \rangle|.
\end{aligned}
$$

Thus $|\langle u, v \rangle| < 1$ as required. $\qquad\square$

**Exercise 2.2.9.** In the above proof, we used the fact that $f \neq 0$ implies $\|f\| \neq 0$. Try to come up with a proof for the Cauchy-Schwarz inequality that only uses properties 1, 3 and 4 of inner product (see Definition 2.2.1). We will have an occasion to use this.

**Theorem 2.2.10** (The triangle inequality). *For all $f, g \in G$,*

$$
\|f + g\| \leq \|f\| + \|g\|.
$$

*Equality holds if and only if one of the vectors is a nonnegative multiple of the other.*

*Proof.* One expands $\|f + g\|^2$ and then uses Cauchy-Schwarz:

$$
\begin{aligned}
\|f + g\|^2 &= \|f\|^2 + 2\operatorname{Re}\langle f, g \rangle + \|g\|^2 \\
&\leq \|f\|^2 + 2|\langle f, g \rangle| + \|g\|^2 \\
&\leq \|f\|^2 + 2\|f\|\|g\| + \|g\|^2 \\
&= (\|f\| + \|g\|)^2.
\end{aligned}
$$

By Theorem 2.2.8 equality occurs in the second inequality if and only if one vector is a constant multiple of the other; equality then occurs in the first inequality if and only if the constant is nonnegative. $\qquad\square$

We see that $\| \cdot \|$ satisfies all the familiar properties of a *norm*:

1. $\|f\| \geq 0$ for all $f \in G$, and $\|f\| = 0$ implies $f = 0$.

2. $\|af\| = |a|\|f\|$ for all $f \in G$ and all $a \in \mathbb{C}$.

3. $\|f + g\| \leq \|f\| + \|g\|$ for all $f, g \in G$.

It follows that $\| \cdot \|$ induces a metric $d$ on $G$ by

$$
d(f, g) = \|f - g\|.
$$

**Exercise 2.2.11.** Prove that the function $d : G \times G \to [0, \infty)$ defined above is a metric.

Thus, every inner product space $G$ carries the structure of a metric space, and the structure of the metric space is related to the linear structure through the inner product.

**Definition 2.2.12.** If $\{f_n\}_{n=0}^\infty$ is a sequence in $G$, we say that $f_n$ *converges to* $f$, and we write $f_n \to f$, if

$$\lim_{n\to\infty} d(f_n, f) = \lim_{n\to\infty} \|f_n - f\| = 0.$$

The following three exercises are easy, but important. We will use them throughout without mention.

**Exercise 2.2.13.** Prove that the inner product is continuous in its variables. That is, show that if $f_n \to f$ and $g_n \to g$, then

$$\langle f_n, g_n \rangle \to \langle f, g \rangle.$$

In particular, the norm is continuous.

**Exercise 2.2.14.** Prove that the vector space operations are continuous. That is, show that if $f_n \to f$ and $g_n \to g$ in $G$ and $c_n \to c$ in $\mathbb{C}$, then

$$f_n + c_n g_n \to f + cg.$$

**Exercise 2.2.15.** Let $G$ be an inner product space, let $g \in G$, and let $F$ be a dense subset of $G$. Show that if $\langle f, g \rangle = 0$ for all $f \in F$, then $g = 0$.

## 2.2.2 Hilbert spaces

Recall that a sequence $\{f_n\}_{n=1}^\infty$ in $G$ is said to be a **Cauchy sequence** if $d(f_n, f_m) = \|f_n - f_m\| \to 0$ as $m, n \to \infty$, and that $G$ is said to be **complete** if every Cauchy sequence actually converges to a point in $G$ (see Section A.2).

**Definition 2.2.16.** An inner product space $G$ is said to be a **Hilbert space** if it is complete with respect to the metric induced by the norm.

Let us review the examples we saw above.

**Example 2.2.17.** The space $G = \mathbb{C}^n$ is complete with respect to the Euclidean norm, so it is a Hilbert space. (In case the reader has not yet seen the completeness of $\mathbb{C}^n = \mathbb{R}^{2n}$, a proof can be adapted from Example 2.2.19).

**Example 2.2.18.** The space $G = C([0, 1])$ is not complete with respect to the norm corresponding to the inner product specified in Example 2.2.4, so this inner product space is not a Hilbert space. Indeed, let $f_n$ be the continuous function that is equal to zero between 0 and $1/2$, equal to 1 between $1/2 + 1/n$ and 1, and linear on $[1/2, 1/2 + 1/n]$. If $m < n$, then $\|f_m - f_n\| < 1/m$, thus this series is Cauchy. However, one can show that if $f$ is a continuous function such that $\|f_n - f\| \to 0$, then $f|_{(0,1/2)} \equiv 0$ and $f|_{(1/2,1)} \equiv 1$. But there is no such continuous function, so $C([0, 1])$ is not complete.

**Example 2.2.19.** Let $\ell^2(\mathbb{N})$ (or, for short, $\ell^2$) denote the set of all square summable sequences of complex numbers. That is, let $\mathbb{C}^{\mathbb{N}}$ denote the space of all complex valued sequences, and define

$$\ell^2 = \left\{ x = (x_k)_{k=0}^{\infty} \in \mathbb{C}^{\mathbb{N}} : \sum_{k=0}^{\infty} |x_k|^2 < \infty \right\}.$$

We endow $\ell^2$ with the natural vector space operations, and with the inner product $\langle x, y \rangle = \sum_k x_k \overline{y}_k$. The norm is given by $\|x\| = \sqrt{\sum |x_k|^2}$.

There are several things to prove in order to show that $\ell^2$ is a Hilbert space. First, we have to show that $\ell^2$ is a vector space; second, we must show that our definition of the inner product makes sense; third, we have to show that $\ell^2$ is complete.

The only nontrivial thing to check with regard to $\ell^2$ being a vector space is that if $x, y \in \ell^2$, then $x + y \in \ell^2$. Using the triangle inequality in $\mathbb{C}^{n+1}$ we find

$$\sum_{k=0}^{n} |x_k + y_k|^2 \le \left( \sqrt{\sum_{k=0}^{n} |x_k|^2} + \sqrt{\sum_{k=0}^{n} |y_k|^2} \right)^2 \le (\|x\| + \|y\|)^2 < \infty$$

for all $n$. Letting $n \to \infty$ we see that $x + y \in \ell^2$. In a similar manner, one uses the Cauchy-Schwarz inequality in $\mathbb{C}^n$ to verify that $\sum x_k \overline{y}_k$ converges absolutely for all $x, y \in \ell^2$, and thus the inner product is well-defined. Checking that the map $(x, y) \mapsto \langle x, y \rangle$ defines an inner product is now straightforward.

We turn to prove that $\ell^2$ is complete. Suppose that $\{x^{(n)}\}_{n=1}^{\infty}$ is a Cauchy sequence in $\ell^2$. That means that for every $n$, $x^{(n)} = (x_k^{(n)})_{k=0}^{\infty}$ is a sequence in $\ell^2$ and that

$$\|x^{(n)} - x^{(m)}\|^2 = \sum_k |x_k^{(n)} - x_k^{(m)}|^2 \longrightarrow 0$$

as $m, n \to \infty$. In particular, for all $k$, we have that $|x_k^{(n)} - x_k^{(m)}|^2 \longrightarrow 0$, thus for every fixed $k$ the sequence $\{x_k^{(n)}\}_{n=1}^{\infty}$ is a Cauchy sequence of complex numbers. By completeness of $\mathbb{C}$, for each $k$ there is some $x_k$ such that $\{x_k^{(n)}\}_{n=1}^{\infty}$ converges to $x_k$. Define $x = (x_k)_{k=0}^{\infty}$. We need to show that $x \in \ell^2$, and that $\{x^{(n)}\}_{n=1}^{\infty}$ converges to $x$ in the norm of $\ell^2$.

Every Cauchy sequence in any metric space is bounded, so there is an $M$ such that $\|x^{(n)}\| \le M$ for all $n$. Fix some $\epsilon > 0$, and choose an integer $n_0$ such that $\|x^{(n)} - x^{(m)}\| < \epsilon$ for all $m, n \ge n_0$. Then for every integer $N$,

$$\sum_{k=0}^{N} |x_k^{(n)} - x_k^{(m)}|^2 < \epsilon^2$$

for all $m, n \ge n_0$. Letting $m \to \infty$, we find that

$$\sum_{k=0}^{N} |x_k^{(n)} - x_k|^2 \le \epsilon^2$$

for all $n \geq n_0$, and therefore $\sum_{k=0}^{N} |x_k|^2 \leq (M+\epsilon)^2$. Since $N$ can be arbitrarily large while $M$ and $\epsilon$ are held fixed, we conclude that $x \in \ell^2$ and that $\|x^{(n)} - x\| \leq \epsilon$ for all $n \geq n_0$. Since $\epsilon$ was arbitrary, we also get that $x^{(n)} \to x$. That completes the proof that $\ell^2$ is a Hilbert space.

It is convenient at times to consider the close relatives of $\ell^2$ defined as follows.

**Example 2.2.20.** Let $S$ be a set. Then we define $\ell^2(S)$ to be

$$\ell^2(S) = \left\{ f : S \to \mathbb{C} : \sum_{s \in S} |f(s)|^2 < \infty \right\}$$

with the inner product $\langle f, g \rangle = \sum_{s \in S} f(s)\overline{g(s)}$. For example, there is the very useful space $\ell^2(\mathbb{Z})$. In case $S$ is uncountable, then one has to be careful how this sum is defined. This issue will be discussed later on (see Definition 3.3.4).

---

## 2.3 Completion

**Theorem 2.3.1.** *Let $G$ be an inner product space. There exists a Hilbert space $H$, and a linear map $V : G \to H$ such that*

1. *For all $g, h \in G$, $\langle V(f), V(g) \rangle_H = \langle f, g \rangle_G$.*

2. *$V(G)$ is dense in $H$.*

*If $H'$ is another Hilbert space and $V' : G \to H'$ is a linear map satisfying the above two conditions, then there is a bijective linear map $U : H \to H'$ such that $\langle U(h), U(k) \rangle_{H'} = \langle h, k \rangle_H$ for all $h, k \in H$ and $U(V(g)) = V'(g)$ for all $g \in G$.*

**Remarks 2.3.2.** Before the proof we make the following remarks.

1. The Hilbert space $H$ is said to be **the Hilbert space completion** or simply **the completion** of $G$.

2. The last assertion of the theorem says that $H$ is essentially unique, up to the obvious "isomorphism" appropriate for Hilbert spaces.

3. Since $V$ preserves all of the structure that $G$ enjoys, one usually identifies $G$ with $V(G)$, and then the theorem is stated as follows: *There exists a unique Hilbert space $H$ which contains $G$ as a dense subspace.*

*Proof.* For the moment being, forget that $G$ is an inner product space, and let $H$ be the metric space completion of $G$, when $G$ is considered simply as a metric space (see Theorem A.2.10). Let $V$ denote the isometric embedding of $G$ into $H$. Thus $H$ is a complete metric space and $V(G)$ is dense in $H$.

It remains to give $H$ the structure of a Hilbert space, to show that the metric is induced by the inner product, and to show that $V$ is a linear map that preserves the inner product. The main idea of the rest of the proof is roughly as follows: every uniformly continuous map from a metric space $X$ into a complete metric space $Z$ extends uniquely to a continuous map from the completion of $X$ to $Z$ (see Exercise A.2.15), thus the vector space operations and the inner product of $G$ extend to $H$ to make it a Hilbert space.

Now to the details. Since $V$ is injective, we can make $V(G)$ into an inner product space by defining vector space operations

$$cV(g) = V(cg),$$

$$V(f) + V(g) = V(f + g),$$

and an inner product

$$\langle V(f), V(g) \rangle = \langle f, g \rangle,$$

for $f, g \in G$ and $c \in \mathbb{C}$. To simplify notation, let us identify $G$ with $V(G)$, and omit $V$. We can do this because $V$ is injective, and because $V$ is compatible with the only structure that we initially had on $V(G)$, which is the metric. Once we extend the linear and inner product structure from $V(G)$ to $H$, the map $V$ will then be a linear map preserving inner products, as required.

Given $h, k \in H$, and letting $\{f_n\}_{n=1}^{\infty}, \{g_n\}_{n=1}^{\infty}$ be sequences in $G$ such that $f_n \to h$ and $g_n \to k$ (in the metric of $H$), we define

$$h + k = \lim_{n \to \infty} f_n + g_n.$$

The limit does exist because $\{f_n + g_n\}_{n=1}^{\infty}$ is a Cauchy sequence in $G$, and $H$ is complete. The limit is also well-defined: if $f_n' \to h, g_n' \to k$, then $d_H(f_n, f_n') \to 0$ and $d_H(g_n, g_n') \to 0$, thus

$$d_H(f_n + g_n, f_n' + g_n') = \|f_n + g_n - (f_n' + g_n')\|_G \to 0.$$

Multiplication of elements in $H$ by scalars is defined in a similar manner, and it follows readily that the operations satisfy the axioms of a vector space.

Next, we define an inner product on $H$. If $h, k \in H$, $G \ni f_n \to h$ and $G \ni g_n \to k$, then we define

$$\langle h, k \rangle_H = \lim_{n \to \infty} \langle f_n, g_n \rangle_G.$$

The limit really does exist: every Cauchy sequence is bounded, so the identity

$$\langle f_n, g_n \rangle - \langle f_m, g_m \rangle = \langle f_n, g_n - g_m \rangle + \langle f_n - f_m, g_m \rangle \qquad (2.1)$$

together with Cauchy-Schwarz implies that $\{\langle f_n, g_n \rangle\}_{n=1}^{\infty}$ is a Cauchy sequence of complex numbers. This implies not only that the limit exists, but that it is independent of the choice of sequences $f_n, g_n$ (why?).

It is easy to check that the form $\langle \cdot, \cdot \rangle : H \times H \to \mathbb{C}$ satisfies the conditions of Definition 2.2.1. For example, if $h \in H$, let $G \ni f_n \to h$. Then $\langle h, h \rangle = \lim \langle f_n, f_n \rangle \geq 0$. Moreover, if $\langle h, h \rangle_H = 0$, then $\|f_n - 0\|^2 = \langle f_n, f_n \rangle \to 0$. Thus $f_n$ converges to 0 in $G \subseteq H$. But $f_n$ converges to $h$ in $H$, therefore, by uniqueness of the limit in $H$, $h = 0$.

By this point, we know that $H$ is a complete metric space which also has the structure of an inner product space. To show that $H$ is a Hilbert space, we need to show that the inner product induces the metric of $H$. Letting $h, k, f_n, g_n$ as above, we have

$$\|h - k\|_H^2 = \langle h - k, h - k \rangle_H = \lim_{n \to \infty} \|f_n - g_n\|_G^2.$$

But since the metric in any metric space is continuous, we also have

$$d_H(h, k) = \lim_{n \to \infty} d_G(f_n, g_n) = \lim_{n \to \infty} \|f_n - g_n\|.$$

Therefore the inner product on $H$ induces a complete metric, as required.

We now come to the final assertion. Let $(H', V')$ be another pair consisting of a Hilbert space and a linear map $V' : G \to H'$ as described in the theorem. We define a map $U_0 : V(G) \to V'(G)$ by $U_0(V(g)) = V'(g)$. From the properties of $V, V'$, it follows that $U_0$ is a linear map, and that

$$\langle U_0(V(f)), U_0(V(g)) \rangle_{H'} = \langle f, g \rangle_G = \langle V(f), V(g) \rangle_H.$$

We have that $U_0$ is a metric preserving map between dense subsets of two complete metric spaces, and thus it extends uniquely to a map $U : H \to H'$. We leave to the reader to show that $U$ is a linear, surjective, inner product preserving map. That completes the proof. $\square$

**Exercise 2.3.3.** Show that the map $U : H \to H'$ is a linear, surjective, inner product preserving map.

## 2.4 The space of Lebesgue square integrable functions

Let $[a, b]$ be a closed interval. Recall that a function $f : [a, b] \to \mathbb{R}$ is said to be *piecewise continuous* if $f$ is continuous at all but finitely many points in $[a, b]$, and if both the right and left one-sided limits exist and are finite at all points in $[a, b]$. A complex-valued function $f$ is said to be piecewise continuous under exactly the same terms. Equivalently, $f$ is piecewise continuous if and only if $\mathrm{Re} f$ and $\mathrm{Im} f$ are both piecewise continuous.

Let $PC[a, b]$ denote the space of piecewise continuous complex-valued functions on the interval $[a, b]$, and define the following form

$$\langle f, g \rangle = \int_a^b f(t)\overline{g(t)}dt. \qquad (2.2)$$

The integral we require is the standard (complex-valued) Riemann integral. Note that this is not exactly an inner product on $PC[a, b]$ because it may happen that $\langle f, f \rangle = 0$ for a function $f$ which is not identically zero; for example, if $f(x) \neq 0$ at finitely many points and zero elsewhere. However, this problem is easily fixed, because of the following fact.

**Exercise 2.4.1.** Let $f \in PC[a, b]$. Prove that $\|f\| = 0$ if and only if the set of points where $f$ is nonzero is a finite set.

To fix the problem, say that $f \sim g$ if $f$ and $g$ are equal at all points of $[a, b]$ except finitely many. Denoting by $\dot{f}$ and $\dot{g}$ the equivalence classes of $f$ and $g$, respectively, we define $\dot{f} + c\dot{g}$ to be the equivalence class of $f + cg$. Clearly, if $f_1 \sim f_2$ and $g_1 \sim g_2$, then $f_1 + cg_1 \sim f_2 + cg_2$ (for $c \in \mathbb{C}$), so the set of equivalence classes becomes a complex vector space. The set of equivalence classes may then be given the structure of an inner product space, with

$$\langle \dot{f}, \dot{g} \rangle = \int_a^b f(t)\overline{g(t)}dt.$$

It is customary to let $PC[a, b]$ denote also the set of equivalence classes of piecewise continuous functions, and to consider elements of this space as functions, with the provision of declaring two functions to be *equal* if they are equivalent. In other words, $PC[a, b]$ can be considered as the space of piecewise continuous functions with the inner product (2.2), where one does not distinguish between two functions if their values differ only on a finite set of points.

The space $C([a, b])$ of all continuous complex-valued functions on the interval $[a, b]$ is contained in the space of piecewise continuous functions, but we should be careful because equality in $PC[a, b]$ is defined modulo finite sets.

**Exercise 2.4.2.** Show that for every $f \in PC[a, b]$, there is at most one $g \in C([a, b])$ in its equivalence class.

The meaning of the exercise above is that our new equality in $PC[a, b]$ is consistent with the equality of functions in $C([a, b])$. Thus, it is safe to say that $C([a, b]) \subseteq PC[a, b]$.

**Exercise 2.4.3.** Justify the statement "$C([a, b])$ is a subspace of the inner product space $PC[a, b]$". Prove that $C([a, b])$ is dense in $PC[a, b]$.

**Exercise 2.4.4.** Show that with respect to the above inner product $PC[a, b]$ is not complete.

In many problems in classical analysis or mathematical physics, such as Fourier series, differential equations, or integral equations, it seems that the natural space of interest is $C([a, b])$ or $PC[a, b]$. Experience has led mathematicians to feel that it is helpful to introduce on these spaces the inner product

$$\langle f, g \rangle = \int_a^b f(t)\overline{g(t)}dt$$

and to use the induced norm as a measure of size (or distance) in this space. However, neither of these spaces is complete. The completion of either of these spaces is denoted $L^2[a, b]$.

**Definition 2.4.5.** The Hilbert space $L^2[a, b]$ is defined to be the unique completion (given by Theorem 2.3.1) of the inner product space $PC[a, b]$, when the latter is equipped with the inner product (2.2).

**Exercise 2.4.6.** Prove that $C([a, b])$ is dense in $L^2[a, b]$, and hence deduce that $L^2[a, b]$ is also "equal" to the abstract completion of $C([a, b])$. As a first step, clarify to yourself in what sense $C([a, b])$ is "contained in" $L^2[a, b]$. This involves recalling in what sense $C([a, b])$ is "contained in" $PC[a, b]$.

Now, the definition of $L^2[a, b]$ obtained by applying Theorem 2.3.1 is rather abstract: it is a space equivalence classes of Cauchy sequences in $PC[a, b]$. We will soon see that it is possible to think of it as a space of functions but, before that, let us make a side remark and mention that the way we have defined $L^2[a, b]$ here is somewhat unusual.

**Remark 2.4.7.** The space $L^2[a, b]$ can be defined in a completely different manner, in rather concrete function-theoretic terms, as the space of *square integrable Lebesgue measurable functions* on the interval (to be precise, equivalence classes of square integrable Lebesgue measurable functions). This is the way this space is defined in a course in measure theory, and is in fact how it is usually defined; we refer the reader to either [11] or [29] for the standard treatment. After one defines the space $L^2[a, b]$, one can define the inner product on it using the *Lebesgue integral.* [1] One can then prove that the inner product space $L^2[a, b]$ is complete, and that the continuous functions (or the piecewise continuous functions) are dense in this space. However, the uniqueness part of Theorem 2.3.1 guarantees that we get precisely the same Hilbert space as we do from the construction we carried out above.

Note that our functional-analytic construction of $L^2[a, b]$ does not give any measure theoretic result for free; it is the hard-analytic theorems in measure theory (completeness of $L^2[a, b]$ and density of the piecewise continuous functions in it) that allow us to conclude that the measure theoretic construction and the abstract construction give rise to the "same" Hilbert space. The way

---

[1]The Lebesgue integral is an extension of the Riemann integral that agrees with the Riemann integral on Riemann-integrable functions, but is defined also for much larger class of functions.

we constructed $L^2[a, b]$ is a shortcut, and students who are serious about their mathematical training should eventually learn the measure theoretic construction.

In this book we do not require any knowledge from measure theory. Rather, we will stick with our definition of $L^2[a, b]$ as an abstract completion, and we will derive some of its function theoretic nature from this. This approach is good enough for many applications of the theory.

Let us decide that we will call every element $f \in L^2[a, b]$ a *function*. Since $PC[a, b] \subset L^2[a, b]$, it is clear that we will be identifying some functions — we were already identifying functions at the level of $PC[a, b]$ — so the word *function* has a somewhat different meaning than what one may be used to. If $f \in L^2[a, b]$ we cannot really say what is the value of $f$ at a point $x \in [a, b]$ (already in $PC[a, b]$ we could not), but $f$ has some other function-like aspects.

First of all, $f$ is "square integrable" on $[a, b]$. To be precise, we define the integral of $|f|^2$ on $[a, b]$ to be

$$\int_a^b |f(t)|^2 dt = \langle f, f \rangle.$$

We see that if $f$ and $g$ are functions in $L^2[a, b]$, then we consider $f$ and $g$ as being equal if $\int_a^b |f(t) - g(t)|^2 dt = 0$.

Second, every $f \in L^2[a, b]$ is actually "integrable" on $[a, b]$, that is, we can define

$$\int_a^b f(t)dt = \langle f, 1 \rangle,$$

where 1 is the constant function 1 on $[a, b]$, which is in $PC[a, b]$, and therefore in $L^2[a, b]$.

Lastly, we can define the integral of $f$ on every sub-interval $[c, d] \subseteq [a, b]$:

$$\int_c^d f(t)dt = \langle f, \chi_{[c,d]} \rangle,$$

where $\chi_{[c,d]}$ is the characteristic function of $[c, d]$, that is, the piecewise continuous function that is equal to 1 on $[c, d]$ and zero elsewhere.

All these definitions are consistent with the definitions of the integral of piecewise continuous functions. Moreover, for a continuous function $f$, if we know $\int_c^d f(t)dt$ for every interval $[c, d] \subseteq [a, b]$ then we can completely recover $f$; for a piecewise continuous function $f$, if we know $\int_c^d f(t)dt$ for every interval $[c, d] \subseteq [a, b]$ then we can recover the equivalence class of $f$ in $PC[a, b]$. It follows (with some work) that if $f \in L^2[a, b]$ then the collection of quantities $\langle f, 1_{[c,d]} \rangle$ for all $c, d$ determines $f$, and consequently we can say that a function in $L^2[a, b]$ is uniquely determined by its integrals over all intervals.

**Exercise 2.4.8.** Prove that a function in $C([a, b])$ is completely determined by the values of its integrals over sub-intervals of $[a, b]$, in the sense that if $f, g \in C([a, b])$, and if $\int_c^d f = \int_c^d g$ for all $a \leq c < d \leq b$, then $f = g$.

**Exercise 2.4.9.** Show that a function in $L^2[a, b]$ is completely determined by the values of its integrals over sub-intervals of $[a, b]$.

You are meant to prove the above exercise using only our definition of $L^2[a, b]$ as the abstract completion of $PC[a, b]$ and the subsequent definition of integral over a sub-interval. It is also a fact in measure theory, that any "function" $f \in L^2[a, b]$ is nothing more and nothing less than the totality of values $\int_c^d f(t)dt$.

**Remark 2.4.10.** When trying to think of $L^2[a, b]$ as a space of functions, there are two questions to ponder. First, given an element $f \in L^2[a, b]$, in what sense can it be considered as a function? We addressed this issue above. Second, given a function $f : [a, b] \to \mathbb{C}$, we ask whether or not $f$ can be considered as an element in $L^2[a, b]$. This is an important question, but for our purposes it is not crucial to settle it now. We will return to this question in Section 12.1. For a full understanding of $L^2[a, b]$, the student is advised to learn the theory of Lebesgue measure and integration.

**Exercise 2.4.11** (Tricky). Try to justify the statement: for every $r < 1/2$, the function given by $f(x) = x^{-r}$ for $x \in (0, 1]$ (and undefined for $x = 0$) is in $L^2[0, 1]$. As a first step, you have to *decide*, and then explain, precisely what it means for $f$ to be in $L^2[a, b]$. Can you say anything about the containment of $f$ in $L^2[a, b]$ if $r \geq 1/2$ and $f$ is given by the same formula? (We will return to this delicate issue in Section 12.1 below).

In a way similar to the above, if $K$ is a nice subset of $\mathbb{R}^k$, then we define $L^2(K)$ to be the completion of the space $C(K)$ with respect to the inner product

$$\langle f, g \rangle = \int_K f(t_1, \ldots, t_k)\overline{g(t_1, \ldots, t_k)}dt_1 \cdots dt_k.$$

The details are similar to the one-dimensional case and we do not dwell on them. $L^2$ spaces can be defined on spaces of a more general nature, but that is best done with some measure theory (again we refer the reader to either [11] or [29]).

## 2.5 Additional exercises

**Exercise 2.5.1.** Let $G$ be an inner product space. Prove that for every $g_1, \ldots, g_n \in G$, there exist scalars $c_1, \ldots, c_n$ with $|c_i| = 1$ for all $i$, such that

$$\left\| \sum_{i=1}^n c_i g_i \right\|^2 \geq \sum_{i=1}^n \|g_i\|^2.$$

Is there anything to say about what happens when the scalar field varies?

**Exercise 2.5.2.** Let $H$ be a Hilbert space, and let $\{h_n\}_{n=1}^{\infty} \subset H$ be a sequence for which there is a constant $C$, such that for every sequence $\{c_n\}_{n=1}^{N}$ of scalars, $|c_n| \leq 1$ for all $n = 1, \ldots, N$,

$$\left\| \sum_{n=1}^{N} c_n h_n \right\| \leq C.$$

Prove that the limit

$$\lim_{N \to \infty} \sum_{n=1}^{N} h_n$$

exists in norm.

**Exercise 2.5.3.** Prove that the inner product space $C([0,1])$ with the inner product specified in Example 2.2.4 is not complete (in other words, provide the details missing from Example 2.2.18).

**Exercise 2.5.4.** Consider the form

$$\langle f, g \rangle = \sum_{k=0}^{\infty} f^{(k)}(0) g^{(k)}(0).$$

where $f^{(k)}(0)$ denotes the $k$th derivative of a function $f$ at 0. Which of the following spaces is an inner product space, when equipped with the above form? Which of them is a Hilbert space?

1. The space

$$H_1 = \{f : (-1,1) \to \mathbb{R} : f \text{ is differentiable infinitely many times}\}.$$

2. The space

$$H_2 = \left\{ f \in H_1 : \sum_{k=0}^{\infty} |f^{(k)}(0)|^2 < \infty \right\}.$$

3. The space

$$H_3 = \{f : \mathbb{D} \to \mathbb{C} : f \text{ is bounded and analytic}\}.$$

4. The space

$$H_4 = \left\{ f : \mathbb{D} \to \mathbb{C} : f \text{ is analytic and } \sum_{k=0}^{\infty} |f^{(k)}(0)|^2 < \infty \right\}.$$

**Exercise 2.5.5.** Let $[\cdot, \cdot] : \mathbb{C}^n \times \mathbb{C}^n \to \mathbb{C}$. Prove that $[\cdot, \cdot]$ is an inner product on $\mathbb{C}^n$ if and only if there exists a positive definite matrix $A \in M_n(\mathbb{C})$ such that

$$[x, y] = \langle Ax, y \rangle,$$

where $\langle \cdot, \cdot \rangle$ denotes the standard inner product on $\mathbb{C}^n$. Likewise, characterize real-valued inner products on the real space $\mathbb{R}^n$.

**Exercise 2.5.6.** Define a norm $\| \cdot \|$ on $\mathbb{R}^2$ by $\|(x_1, x_2)\| = \max\{|x_1|, |x_2|\}$. Can you find an inner product on $\mathbb{R}^2$ such that $\| \cdot \|$ is the norm induced by this inner product?

**Exercise 2.5.7.** Can one define an inner product on $C(X)$ which induces the supremum norm?

**Exercise 2.5.8.** The *Bergman space* $L_a^2(\mathbb{D})$ is the space of all analytic functions on the unit disc that are square integrable, that is

$$L_a^2(\mathbb{D}) = \left\{ f : \mathbb{D} \to \mathbb{C} \text{ analytic } : \lim_{r \nearrow 1} \int_{x^2 + y^2 \leq r} |f(x + iy)|^2 dx dy < \infty \right\}.$$

On $L_a^2(\mathbb{D})$, define the inner product

$$\langle f, g \rangle = \int_{x^2 + y^2 < 1} f(x + iy) \overline{g(x + iy)} dx dy.$$

Prove that $L_a^2(\mathbb{D})$ is a Hilbert space (this will be established in Section 6.2).

**Exercise 2.5.9.** Prove that each one of the following two sets is dense in $L^2[0, 1]$.

1. The space of step functions functions $\operatorname{span}\{\chi_{[a,b]} : a, b \in [0, 1]\}$.

2. The space of polynomials.

**Exercise 2.5.10.** Let $\{H_n\}_{n \in \mathbb{N}}$ be a family of Hilbert spaces. Define a new space $\oplus_{n \in \mathbb{N}} H_n$ as follows:

$$\bigoplus_{n \in \mathbb{N}} H_n = \{(h_n)_{n=0}^\infty : h_n \in H_n \text{ for all } n \in \mathbb{N}, \sum_{n=0}^\infty \|h_n\|_{H_n}^2 < \infty\}.$$

On $\oplus_{n \in \mathbb{N}} H_n$ define an inner product

$$\left\langle (g_n)_{n=0}^\infty, (h_n)_{n=0}^\infty \right\rangle = \sum_{n=0}^\infty \langle g_n, h_n \rangle_{H_n}.$$

Prove that this inner product is well-defined, and that $\oplus_{n \in \mathbb{N}} H_n$ is a Hilbert space.

The space $\oplus_{n \in \mathbb{N}} H_n$ is called the **direct sum** of the Hilbert spaces $\{H_n\}_{n \in \mathbb{N}}$. Note that every space $H_n$ can be considered to be a closed subspace of $\oplus_{n \in \mathbb{N}} H_n$. Sometimes it is called the **external direct sum**, to make a distinction from the situation where all the spaces $H_n$ are a priori mutually orthogonal subspaces inside a given Hilbert space. When one has a finite family $H_1, \ldots, H_N$, then one may form the external direct sum $\oplus_{n=1}^N H_n$, denoted also $H_1 \oplus \cdots \oplus H_N$, in a similar manner. The direct sum of a not necessarily countable family of Hilbert spaces can also be defined (see Exercise 3.6.8).

# Chapter 3

## Orthogonality, projections, and bases

### 3.1 Orthogonality

**Definition 3.1.1.** Let $G$ be an inner product space.

1. Two vectors $f, g \in G$ are said to be **orthogonal**, denoted $f \perp g$, if $\langle f, g \rangle = 0$.

2. A set of nonzero vectors $\{e_i\}_{i \in I} \subseteq G$ is said to be an **orthogonal** set if $e_i \perp e_j$ for all $i \neq j$.

3. An orthogonal set $\{e_i\}_{i \in I} \subseteq G$ is said to be an **orthonormal** set if $e_i$ is a unit vector (i.e., $\|e_i\| = 1$) for all $i \in I$. An orthonormal set is sometimes called an **orthonormal system**.

The following two easy propositions show how the geometry of inner product spaces has some close similarities with Euclidean geometry.

**Proposition 3.1.2** (Pythagorean identity). *In an inner product space, the following hold*

1. *If $f \perp g$, then $\|f + g\|^2 = \|f\|^2 + \|g\|^2$.*

2. *If $\{e_i\}_{i=1}^{n}$ is a finite orthogonal set, then*

$$\left\| \sum_i e_i \right\|^2 = \sum_i \|e_i\|^2.$$

*Proof.* The first assertion is a special case of the second, which in turn follows from

$$\left\| \sum_i e_i \right\|^2 = \left\langle \sum_i e_i, \sum_j e_j \right\rangle = \sum_{i,j=1}^{n} \langle e_i, e_j \rangle = \sum_i \|e_i\|^2.$$

$\square$

**Example 3.1.3.** Let $G = C([0,1])$ with the usual inner product. The set $\{e^{2\pi i n x}\}_{n=-\infty}^{\infty}$ is an orthonormal set in $G$. Indeed,

$$\langle e^{2\pi i m x}, e^{2\pi i n x} \rangle = \int_0^1 e^{2\pi i (m-n)x} dx,$$

and this is equal to 1 if $m = n$ and to $\frac{1}{2\pi i(m-n)} e^{2\pi i(m-n)x} \Big|_0^1 = 0$ if $m \neq n$.
The set $\{1, \sin 2\pi nx, \cos 2\pi nx\}_{n=1}^{\infty}$ is an orthogonal set, but not orthonormal. These two systems are also orthogonal sets in the larger space $H = L^2[0, 1]$.

Another immediate identity that holds in inner product spaces is the following.

**Proposition 3.1.4** (The parallelogram law). *For any $f, g$ in an inner product space, the following identity holds:*

$$\|f + g\|^2 + \|f - g\|^2 = 2\|f\|^2 + 2\|g\|^2.$$

**Exercise 3.1.5.** Prove the parallelogram law.

The parallelogram law differs from the Pythagorean identity in that it is stated only in terms of the norm, and makes no mention of the inner product. Thus, it can be used to show that some norms are not induced by an inner product (recall Exercise 2.5.7; see also Exercise 7.1.12).

---

## 3.2 Orthogonal projection and orthogonal decomposition

**Definition 3.2.1.** A subset $S$ in a vector space $V$ is said to be **convex** if for all $x, y \in S$ and all $t \in [0, 1]$, the vector $tx + (1 - t)y$ is also in $S$.

**Lemma 3.2.2.** *Let $S$ be a closed convex set in a Hilbert space $H$. Then there is a unique $h \in S$ of minimal norm.*

*Proof.* Put $d = \inf\{\|y\| : y \in S\}$. Let $\{y_n\}$ be a sequence in $S$ such that $\|y_n\| \to d$. Applying the parallelogram law to $\frac{1}{2}y_m$ and $\frac{1}{2}y_n$, we find

$$\left\|\frac{1}{2}(y_m - y_n)\right\|^2 = 2\left\|\frac{1}{2}y_n\right\|^2 + 2\left\|\frac{1}{2}y_m\right\|^2 - \left\|\frac{1}{2}(y_n + y_m)\right\|^2.$$

Now, $\|\frac{1}{2}(y_n + y_m)\|^2 \geq d^2$, thus, letting $m, n \to \infty$, we find that the right-hand side must tend to zero. Hence $\{y_n\}$ is a Cauchy sequence. Since $H$ is complete and $S$ is closed, there is an $h \in S$ such that $y_n \to h$. By the continuity of the norm, $\|h\| = d$. This proves the existence of a norm minimizer.

To prove the uniqueness, let $g, h \in S$ such that $\|g\| = \|h\| = d$. If we form the sequence $g, h, g, h, \ldots$, then we have just seen above that this is a Cauchy sequence. It follows that $g = h$. $\square$

**Theorem 3.2.3.** *Let $S$ be a closed convex set in a Hilbert space $H$, and let $h \in H$. Then there exists a unique $g \in S$ such that*

$$\|g - h\| \leq \|f - h\| \quad \text{for all } f \in S.$$

*Proof.* Apply Lemma 3.2.2 to the convex set $S - h = \{g - h : g \in S\}$. The unique element $y \in S - h$ of minimal norm corresponds to a unique $g \in S$ (given by $g = y + h$) such that $\|g - h\|$ is minimal, so the theorem follows. □

We call the element $g \in S$ **the best approximation for $h$ within $S$**. We let $P_S$ denote the map that assigns to each $h \in H$ the best approximation for $h$ within $S$. Thus, we have $\|P_S(h) - h\| \leq \|f - h\|$ for all $f \in S$. Since every subspace is convex, we immediately obtain the following.

**Corollary 3.2.4.** *Let $M$ be a closed subspace of $H$, and let $h \in H$. Then there exists a unique $g \in M$ such that*

$$\|g - h\| \leq \|m - h\| \quad \text{for all } m \in M.$$

**Theorem 3.2.5.** *Let $S$ be a closed convex set in a Hilbert space $H$, and let $h \in H$ and $g \in S$. The following are equivalent:*

1. *$g = P_S(h)$ (in other words, $g$ is the best approximation for $h$ within $S$).*

2. *$\operatorname{Re}\langle h - g, f - g \rangle \leq 0$ for all $f \in S$.*

**Remark 3.2.6.** Before proceeding with the proof, we pause to get a picture of the geometric meaning of the theorem. Consider the real Hilbert space $\mathbb{R}^2$. Make a picture of a convex set $S$, a point $h$ not in the set, its best approximation $g = P_S(h)$, and another point $f \in S$. Draw a straight line between $g$ and $h$ and another line between $g$ and $f$. Note that the angle between these lines is greater than $\frac{\pi}{2}$. This is the geometric meaning of the condition $\operatorname{Re}\langle h - g, f - g \rangle \leq 0$ above.

*Proof.* Let $g = P_S(h)$ and $f \in S$. Expanding the inequality

$$\|h - (tf + (1 - t)g)\|^2 \geq \|h - g\|^2$$

(which holds for all $t \in (0, 1)$), we get

$$\|h - g\|^2 - 2\operatorname{Re}\langle h - g, t(f - g)\rangle + \|t(f - g)\|^2 \geq \|h - g\|^2.$$

Dividing by $t$ and cancelling some terms, we obtain

$$2\operatorname{Re}\langle h - g, f - g \rangle \leq t\|f - g\|^2.$$

Since this is true for all $t \in (0, 1)$, we conclude that $2\operatorname{Re}\langle h - g, f - g \rangle \leq 0$.

To get the converse implication, let $f \in S$. Then using $2\operatorname{Re}\langle h - g, f - g \rangle \leq 0$ we find

$$\begin{aligned}
\|h - f\|^2 &= \|(h - g) - (f - g)\|^2 \\
&= \|h - g\|^2 - 2\operatorname{Re}\langle h - g, f - g \rangle + \|f - g\|^2 \\
&\geq \|h - g\|^2,
\end{aligned}$$

so $g = P_S(h)$. □

**Corollary 3.2.7.** *Let $M$ be a closed subspace in a Hilbert space $H$, and let $h \in H$ and $g \in M$. The following are equivalent:*

1. $g = P_M(h)$.

2. $h - g \perp m$ *for all* $m \in M$.

**Definition 3.2.8.** *Let $G$ be an inner product space, and let $S \subseteq G$. We define*

$$S^{\perp} = \{g \in G : \langle s, g \rangle = 0 \ \text{ for all } s \in S\}.$$

**Exercise 3.2.9.** *Prove that for any set $S$, $S^{\perp}$ is a closed subspace, and that $S \cap S^{\perp} \subseteq \{0\}$.*

**Theorem 3.2.10.** *Let $M$ be a closed subspace in a Hilbert space $H$. Then for every $h \in H$ there is a unique $m \in M$ and a unique $n \in M^{\perp}$ such that $h = m + n$.*

*Proof.* Write $h = P_M h + (h - P_M h)$. Setting $m = P_M h$, we have that $m \in M$ by definition. Moreover, letting $n = h - m = h - P_M h$, we have $n \in M^{\perp}$ by Corollary 3.2.7.

For uniqueness, assume that $h = m + n = m' + n'$, with $m' \in M$ and $n' \in M^{\perp}$. Then we have

$$M \ni m - m' = n - n' \in M^{\perp},$$

and since $M \cap M^{\perp} = \{0\}$, we have $m = m'$ and $n = n'$. $\qquad\qquad\square$

**Remark 3.2.11.** The conclusion of the theorem is usually expressed shortly by writing $H = M \oplus M^{\perp}$, and one says that $H$ **is the orthogonal direct sum of $M$ and $M^{\perp}$**. The space $M^{\perp}$ is called the **orthogonal complement of $M$ in $H$**.

**Theorem 3.2.12.** *Let $S$ be a closed convex subset in a Hilbert space $H$. The map $P_S$ satisfies $P_S \circ P_S = P_S$ , and $P_S(h) = h$ if and only if $h \in S$. The map $P_S$ is linear if and only if $S$ is a subspace.*

**Exercise 3.2.13.** Prove Theorem 3.2.12.

**Definition 3.2.14.** *A mapping $T$ satisfying $T \circ T = T$ is called a **projection**. $P_S$ is called the **best approximation projection onto** $S$. If $S$ is a subspace, then $P_S$ is also called **the orthogonal projection (of $H$) onto** $S$.*

**Theorem 3.2.15.** *Let $M$ be a subspace of a Hilbert space $H$. Then $M$ is closed if and only if $M = M^{\perp\perp}$.*

*Proof.* We already noted that $M^{\perp\perp}$ is closed (Exercise 3.2.9), so one direction is immediate. Trivially, $M \subseteq M^{\perp\perp}$ (because anything in $M$ is orthogonal to anything that is orthogonal to everything in $M$). So assume that $M$ is closed and that $x \in M^{\perp\perp}$, and write the orthogonal decomposition of $x$ with respect to the decomposition $H = M \oplus M^\perp$, that is $x = y + z$, with $y \in M, z \in M^\perp$. But $y \in M^{\perp\perp}$, so this is also a decomposition of $x$ with respect to $H = M^\perp \oplus M^{\perp\perp}$. However, $x$ already has a decomposition $x = 0 + x$ with respect to $H = M^\perp \oplus M^{\perp\perp}$. The uniqueness clause in Theorem 3.2.10 now implies that $x = y \in M$. $\qquad\square$

**Corollary 3.2.16.** *For any subspace $M \subseteq H$, $M^{\perp\perp} = \overline{M}$.*

*Proof.* $\overline{M}^\perp = M^\perp$, thus $M^{\perp\perp} = \overline{M}^{\perp\perp} = \overline{M}$. $\qquad\square$

Recall from a course in linear algebra that every finite dimensional inner product space has an orthonormal basis (see Section (3.5) in case you have never seen this or forgotten it).

**Exercise 3.2.17.** Prove that a finite dimensional subspace of an inner product space is closed.

**Proposition 3.2.18.** *Let $M$ be a finite dimensional subspace of a Hilbert space $H$, and let $\{e_i\}_{i=1}^N$ be an orthonormal basis for $M$. For all $h \in H$,*

$$P_M h = \sum_{i=1}^N \langle h, e_i \rangle e_i.$$

*Proof.* Put $m = P_M h$. Since $m \in M$, then $m = \sum_{i=1}^N \langle m, e_i \rangle e_i$. (The familiar proof from the course in linear algebra goes as follows:

$$m = \sum_{i=1}^N c_i e_i$$

for some constants $c_i$. Taking the inner product of this equality with $e_k$ one obtains

$$\langle m, e_k \rangle = \sum_{i=1}^N c_i \langle e_i, e_k \rangle = c_k,$$

and the representation of $m$ is as we claimed). But, by Corollary 3.2.7, $h - m$ is orthogonal to $M$. In other words, $\langle h - m, e_i \rangle = 0$, or $\langle h, e_i \rangle = \langle m, e_i \rangle$, for all $i$; therefore, $m = \sum_{i=1}^N \langle h, e_i \rangle e_i$, as asserted. $\qquad\square$

---

## 3.3 Orthonormal bases

Recall that a **Hamel basis** for a vector space $V$ is a family $\{v_i\}_{i \in I}$ such that every $v \in V$ can be written in a unique way as a (finite) linear combination of the $v_i$s (in linear algebra, a Hamel basis is called simply a *basis*, since no other notions of basis are considered). A vector space is said to be **infinite dimensional** if it has no finite Hamel basis.

**Exercise 3.3.1.** Prove that if $H$ is an infinite dimensional Hilbert space, then $H$ has no countable Hamel basis. (If you solved Exercise 1.2.1, then try to give a totally different proof, which is Hilbert space specific.)

We see that Hamel bases are rather inaccessible objects in infinite dimensional Hilbert spaces. We will need another notion of basis.

**Definition 3.3.2.** Let $E = \{e_i\}_{i \in I}$ be an orthonormal system in an inner product space $G$. $E$ is said to be ***complete*** if $E^\perp = \{0\}$.

**Proposition 3.3.3.** *Every inner product space has a complete orthonormal system.*

*Proof.* One considers the set of all orthonormal systems in the space, partially ordered by inclusion, and applies Zorn's lemma to deduce the existence of a maximal orthonormal system. A maximal orthonormal system must be complete, otherwise one could add a normalized perpendicular vector. □

In case that the inner product space in question is separable, one can also prove that there exists a complete orthonormal system by applying the Gram-Schmidt process to a dense sequence (see Section 3.5).

In this section, we will prove that a complete orthonormal system in a Hilbert space behaves very much like an orthonormal basis in a finite dimensional space. To make this precise, we first need to make sense of infinite sums in inner product spaces.

**Definition 3.3.4.** Let $I$ be any set, let $\{v_i\}_{i \in I}$ be a set of elements in an inner product space $G$, and let $g \in G$. We say that the series $\sum_{i \in I} v_i$ ***converges to*** $g$, and we write $\sum_{i \in I} v_i = g$, if for all $\epsilon > 0$, there exists a finite set $F_0 \subseteq I$ such that for every finite set $F \subseteq I$,

$$F_0 \subseteq F \quad \text{implies that} \quad \left\| \sum_{i \in F} v_i - g \right\| < \epsilon.$$

The field of complex numbers $\mathbb{C}$ is an inner product space. In particular, if $\{a_i\}_{i \in I}$ is a set of complex numbers, we say that the series $\sum_{i \in I} a_i$ ***converges to*** $a \in \mathbb{C}$, and we write $\sum_{i \in I} a_i = a$, if for all $\epsilon > 0$, there exists a finite set $F_0$ such that for every finite set $F \subseteq I$,

$$F_0 \subseteq F \quad \text{implies that} \quad \left| \sum_{i \in F} a_i - a \right| < \epsilon.$$

**Remark 3.3.5.** The reader is warned that if $I = \{1, 2, \ldots, \}$ and $\{a_i\}_{i \in I}$ is a sequence of real or complex numbers, the definition we just gave for convergence of the series $\sum_{i \in I} a_i$ is different from the usual definition of convergence of the series $\sum_{i=1}^{\infty} a_i$. Recall that the latter is defined as convergence of the sequence of partial sums $\sum_{i=1}^{N} a_i$ as $N \to \infty$.

**Exercise 3.3.6.** Prove that in an inner product space $G$, a series $\sum_{i\in I} v_i$ converges to $g$ if and only if there exists a countable set $J$ contained in $I$ such that

(a) $v_i = 0$ if $i \in I \setminus J$; and

(b) For any enumeration $J = \{j_1, j_2, \ldots, \}$ of the elements of $J$, the limit $\lim_{n\to\infty} \sum_{n=1}^{N} v_{j_n} = g$ exists in $G$.

**Exercise 3.3.7.** Suppose that $a_i \geq 0$ for all $i$. Prove that $\sum_{i\in I} a_i$ converges and is bounded by $M$ if and only if the set of all finite sums

$$\left\{ \sum_{i\in F} a_i : F \subseteq I \text{ finite} \right\}$$

is bounded by $M$.

**Proposition 3.3.8** (Bessel's inequality). *Let $\{e_i\}_{i\in I}$ be an orthonormal set in an inner product space $G$, and let $g \in G$. Then*

$$\sum_{i\in I} |\langle g, e_i \rangle|^2 \leq \|g\|^2.$$

*Proof.* Let $F \subseteq I$ be finite. By Proposition 3.2.18, $\sum_{i\in F}\langle g, e_i \rangle e_i$ is the orthogonal projection of $g$ onto span$\{e_i\}_{i\in F}$, thus, by Corollary 3.2.7, $\sum_{i\in F}\langle g, e_i \rangle e_i$ is orthogonal to $g - \sum_{i\in F}\langle g, e_i \rangle e_i$. Now,

$$g = \sum_{i\in F}\langle g, e_i \rangle e_i + \left( g - \sum_{i\in F}\langle g, e_i \rangle e_i \right),$$

therefore, by the Pythagorean identity,

$$\|g\|^2 = \| \sum_{i\in F}\langle g, e_i \rangle e_i \|^2 + \|g - \sum_{i\in F}\langle g, e_i \rangle e_i \|^2.$$

thus (by Pythagoras once more) $\sum_{i\in F} |\langle g, e_i \rangle|^2 \leq \|g\|^2$. This holds for all $F$, so (invoking Exercise 3.3.7) the assertion follows. $\square$

**Exercise 3.3.9.** Deduce the Cauchy-Schwarz inequality from Bessel's inequality (this should be one line). Did this involve circular reasoning?

**Definition 3.3.10.** Let $\{e_i\}_{i\in I}$ be an orthonormal system in an inner product space $G$. For every $g \in G$, the scalars $\langle g, e_i \rangle$ are called **the (generalized) Fourier coefficients of** $g$ with respect to $\{e_i\}_{i\in I}$.

By Bessel's inequality combined with Exercises 3.3.6 and 3.3.7, for every $g$, only countably many Fourier coefficients are nonzero. This fact frees us, in the following proposition, to consider only countable orthonormal systems.

**Proposition 3.3.11.** *Let $\{e_n\}_{n=1}^{\infty}$ be an orthonormal system in an inner product space $G$. Then, for every given $g \in G$, the following are equivalent:*

1. *$\sum_{n=1}^{\infty} |\langle g, e_n \rangle|^2 = \|g\|^2$.*

2. *$\sum_{n=1}^{\infty} \langle g, e_n \rangle e_n = g$.*

3. *For all $\epsilon > 0$, there exists an integer $N_0$ and scalars $a_1, \ldots, a_{N_0}$ such that $\|g - \sum_{n=1}^{N_0} a_n e_n\| < \epsilon$.*

**Remark 3.3.12.** The convergence of the series of vectors in (2) is to be interpreted simply as the assertion that $\lim_{N \to \infty} \|\sum_{n=1}^{N} \langle g, e_n \rangle e_n - g\| = 0$. Note that the equivalence (1) $\Leftrightarrow$ (2) implies that this vector valued series converges regardless of the order in which it is summed, and also that this series converges in the sense of Definition 3.3.4.

*Proof.* As in the proof of Bessel's inequality, we find that

$$\|g\|^2 = \sum_{n=1}^{N} |\langle g, e_n \rangle|^2 + \|g - \sum_{n=1}^{N} \langle g, e_n \rangle e_n\|^2,$$

and this implies that (1) and (2) are equivalent. (2) obviously implies (3), because one simply takes $a_n = \langle g, e_n \rangle$.

Assume that (3) holds. Let $\epsilon > 0$ be given. We need to find $N_0$, such that for all $N \geq N_0$, $\|\sum_{n=1}^{N} \langle g, e_n \rangle e_n - g\| < \epsilon$. Let $N_0$ be the $N_0$ from (3) corresponding to $\epsilon$, and let $a_1, \ldots, a_{N_0}$ be the corresponding scalars. For any $N \geq N_0$, the linear combination $\sum_{n=1}^{N_0} a_n e_n$ is in the subspace spanned by $e_1, \ldots, e_N$. Define $M = \text{span}\{e_1, \ldots, e_N\}$. But by Proposition 3.2.18, $P_M g = \sum_{n=1}^{N} \langle g, e_n \rangle e_n$ is the best approximation for $g$ within $M$, therefore

$$\|\sum_{n=1}^{N} \langle g, e_n \rangle e_n - g\| \leq \|\sum_{n=1}^{N_0} a_n e_n - g\| < \epsilon.$$

$\square$

**Proposition 3.3.13.** *Let $\{e_i\}_{i \in I}$ be an orthonormal system in a Hilbert space $H$, and let $\{a_i\}_{i \in I}$ be a set of complex numbers. The series $\sum_{i \in I} a_i e_i$ converges in $H$ if and only if $\sum_{i \in I} |a_i|^2 < \infty$.*

*Proof.* If $a_i \neq 0$ for more than countably many values of $i$, then we know that neither one of the sums converges. Thus, we may assume that $I$ is countable. The case where $I$ is finite being trivial, we assume that $I = \{1, 2, 3, \ldots\}$.

Suppose that $\sum_{i=1}^{\infty} a_i e_i$ converges to some $h \in H$, thus

$$\lim_{n \to \infty} \sum_{i=1}^{n} a_i e_i = h.$$

Taking the inner product of the above with $e_k$, we find that $a_k = \langle h, e_k \rangle$ for all $k = 1, 2, 3, \ldots$. By Bessel's inequality, $\sum_{i=1}^{\infty} |a_i|^2 < \infty$.

Conversely, assume that $\sum_{i=1}^{\infty} |a_i|^2 < \infty$. Define $h_n = \sum_{i=1}^{n} a_i e_i$. Then

$$\|h_m - h_n\|^2 = \sum_{i=m+1}^{n} |a_i|^2,$$

for all $m < n$, and it follows that $\{h_n\}$ is a Cauchy sequence. Let $h = \lim_{n \to \infty} h_n$ be the limit of the sequence $\{h_n\}$. We have that $h = \sum_{i \in I} a_i h_i$, and the proof is complete. □

**Theorem 3.3.14.** *Let $\{e_i\}_{i \in I}$ be a complete orthonormal system in a Hilbert space $H$. Then for every $h \in H$,*

$$h = \sum_{i \in I} \langle h, e_i \rangle e_i, \tag{3.1}$$

*and*

$$\|h\|^2 = \sum_{i \in I} |\langle h, e_i \rangle|^2. \tag{3.2}$$

*Proof.* Fix $h \in H$. By Bessel's inequality, $\sum_{i \in I} |\langle h, e_i \rangle|^2 < \infty$. By Proposition 3.3.13, the series $\sum_{i \in I} \langle h, e_i \rangle e_i$ converges. Put $g = h - \sum_{i \in I} \langle h, e_i \rangle e_i$. Our goal is to show that $g = 0$. Since $\{e_i\}_{i \in I}$ is complete, it suffices to show that $\langle g, e_k \rangle = 0$ for all $k \in I$. By continuity of the inner product,

$$\left\langle \sum_{i \in I} \langle h, e_i \rangle e_i, e_k \right\rangle = \langle h, e_k \rangle \langle e_k, e_k \rangle = \langle h, e_k \rangle,$$

whence $\langle g, e_k \rangle = \langle h, e_k \rangle - \langle h, e_k \rangle = 0$ for all $k$. Thus $g \perp \{e_i\}_{i \in I}$, so $g = 0$. This implies (3.1).

By Proposition 3.3.11, (3.1) and (3.2) are equivalent, and the proof is complete. □

Because of the above theorem, a complete orthonormal system in a Hilbert space is called an **orthonormal basis**. Combining the above theorem with Proposition 3.3.3, we see that every Hilbert space has an orthonormal basis.

An orthonormal system (not necessarily in a Hilbert space) satisfying the conclusions of Theorem 3.3.14 is sometimes said to be a **closed orthonormal system**. Perhaps this is to avoid the usage of the word "basis", since an orthonormal basis is definitely not a basis according to the definition given in linear algebra (see Exercise 3.3.1).

**Remark 3.3.15.** The identity (3.2) in Theorem 3.3.14 is called *Parseval's identity*.

**Corollary 3.3.16** (Generalized Parseval identity). *Let $\{e_i\}_{i\in I}$ be an orthonormal basis for a Hilbert space $H$, and let $g, h \in H$. Then*

$$\langle g, h \rangle = \sum_{i \in I} \langle g, e_i \rangle \overline{\langle h, e_i \rangle}.$$

*Proof.* One may deduce this directly, by plugging in $g = \sum \langle g, e_i \rangle e_i$ and $h = \sum \langle h, e_i \rangle e_i$ in the inner product. As an exercise, the reader should think how this follows from Parseval's identity combined with the *polarization identity*

$$\langle x, y \rangle = \frac{1}{4} \left( \|x + y\|^2 - \|x - y\|^2 + i\|x + iy\|^2 - i\|x - iy\|^2 \right),$$

which is true in $H$ as well as in $\mathbb{C}$.                                            □

We are now able to write down the formula for the orthogonal projection onto a closed subspace which is not necessarily finite dimensional.

**Theorem 3.3.17.** *Let $M$ be a closed subspace in a Hilbert space $H$. Let $\{e_i\}_{i\in I}$ be an orthonormal basis for $M$. Then, for every $h \in H$,*

$$P_M h = \sum_{i \in I} \langle h, e_i \rangle e_i.$$

**Exercise 3.3.18.** Prove Theorem 3.3.17.

---

## 3.4   Dimension and isomorphism

**Theorem 3.4.1.** *Let $\{e_i\}_{i\in I}$ and $\{f_j\}_{j\in J}$ be two orthonormal bases for the same Hilbert space $H$. Then $I$ and $J$ have the same cardinality.*

*Proof.* If one of the index sets is finite, then this result follows from linear algebra. So assume both sets are infinite. For every $i \in I$, let $J_i = \{j \in J : \langle e_i, f_j \rangle \neq 0\}$. Every $j \in J$ belongs to at least one $J_i$, because $\{e_i\}_{i\in I}$ is complete. Therefore $J = \cup_{i\in I} J_i$. But as we noted several times, $|J_i| \leq \aleph_0$. These two facts combine to show that the cardinality of $J$ is less than or equal to the cardinality of $I$. Reversing the roles of $I$ and $J$ we see that they must have equal cardinality.

For completeness, let us see that the case where one set is finite can also be proved using Hilbert space methods. Let $\{e_i\}_{i\in I}$ and $\{f_j\}_{j\in J}$ be two orthonormal bases for $H$, and assume that $|I| < \infty$. Let $F$ be a finite subset of $J$. Using Parseval's identity, switching the order of summation, and then

Bessel's inequality, we obtain

$$|F| = \sum_{j \in F} \|f_j\|^2 = \sum_{j \in F} \sum_{i \in I} |\langle f_j, e_i \rangle|^2$$

$$\leq \sum_{i \in I} \|e_i\|^2 = |I|.$$

Thus $|F| \leq |I|$ for every finite subset $F$ of $J$, so $|J| \leq |I| < \infty$. By symmetry we conclude that $|J| = |I|$. $\qquad\square$

**Definition 3.4.2.** Let $H$ be a Hilbert space. The **dimension of $H$** is defined to be the cardinality of any orthonormal basis for $H$.

**Definition 3.4.3.** Let $G_1, G_2$ be inner product spaces. A **unitary map** (or simply a **unitary**) from $G_1$ to $G_2$ is a bijective linear map $U : G_1 \to G_2$ such that $\langle Ug, Uh \rangle_{G_2} = \langle g, h \rangle_{G_1}$ for all $g, h \in G_1$. Two inner product spaces $G_1, G_2$ are said to be **isomorphic** if there exists a unitary map between them.

**Theorem 3.4.4.** *Two Hilbert spaces are isomorphic if and only if they have the same dimension.*

**Exercise 3.4.5.** Prove Theorem 3.4.4.

**Exercise 3.4.6.** Prove that a Hilbert space is separable if and only if its dimension is less than or equal to $\aleph_0$ (recall that a metric space is said to be separable if it contains a countable dense subset).

**Exercise 3.4.7.** Let $I$ be any set. Define

$$\ell^2(I) = \left\{ (x_i)_{i \in I} \in \mathbb{C}^I : \sum_{i \in I} |x_i|^2 < \infty \right\}.$$

Prove that $\ell^2(I)$, with the inner product $\langle (x_i)_{i \in I}, (y_i)_{i \in I} \rangle = \sum_{i \in I} x_i \bar{y}_i$, is a Hilbert space. Prove also that any Hilbert space is isomorphic to $\ell^2(I)$ for some set $I$.

We have gone through some efforts to treat Hilbert spaces which are of arbitrary dimension. However, in mathematical practice, one rarely encounters (or wishes to encounter) a nonseparable space. Nevertheless, rare things do happen. Occasionally it is useful to have an arbitrarily large Hilbert space for some universal construction to be carried out. There are also some natural examples which arise in analysis.

**Exercise 3.4.8.** Let $G$ be the linear span of all functions of the form $\sum_{k=1}^n a_k e^{i\lambda_k t}$, where $\lambda_k \in \mathbb{R}$. On $G$, we define a form

$$\langle f, g \rangle = \lim_{T \to \infty} \frac{1}{2T} \int_{-T}^{T} f(t) \overline{g(t)} dt.$$

Prove that $G$ is an inner product space. Let $H$ be the completion of $G$. Prove that $H$ is not separable. Find an orthonormal basis for $H$.

## 3.5 The Gram-Schmidt process

For completeness, we now give a version of the Gram-Schmidt orthogonalization process appropriate for sequences of vectors. In the case where the sequence is finite, this is precisely the same procedure studied in a course in linear algebra.

**Theorem 3.5.1.** *Let $v_1, v_2, \ldots$ be a sequence of nonzero vectors in an inner product space $G$. Then, there exists an orthonormal sequence $e_1, e_2, \ldots$ with the same linear span as $v_1, v_2, \ldots$. If the sequence $v_1, v_2, \ldots$ is linearly independent, then for all $n = 1, 2, \ldots$,*

$$\mathrm{span}\{e_1, \ldots, e_n\} = \mathrm{span}\{v_1, \ldots, v_n\}.$$

*Proof.* From every sequence of vectors, one can extract a linearly independent sequence with the same span, so it suffices to prove the second half of the theorem. Assume that $v_1, v_2, \ldots$ is a linearly independent sequence. We prove the claim by induction on $n$. For $n = 1$ we put $e_1 = v_1/\|v_1\|$. Assume that $n > 1$, and that we have constructed an orthonormal sequence $e_1, \ldots, e_{n-1}$ such that

$$\mathrm{span}\{e_1, \ldots, e_{n-1}\} = \mathrm{span}\{v_1, \ldots, v_{n-1}\}.$$

Let $M$ denote the subspace appearing in the above equality, and let $P_M$ be the orthogonal projection onto $M$. Then $v_n$ is not in $M$. Put $u = v_n - P_M v_n$. Then $u \neq 0$, and by Corollary 3.2.7, $u \in M^\perp$. Let $e_n = u/\|u\|$. Then $e_1, \ldots, e_n$ is an orthonormal sequence, and $e_n \in \mathrm{span}\{v_1, \ldots, v_n\}$ by construction. Thus $\mathrm{span}\{e_1, \ldots, e_n\} \subseteq \mathrm{span}\{v_1, \ldots, v_n\}$. But since $e_1 \ldots, e_n$ are $n$ linearly independent vectors, we must have $\mathrm{span}\{e_1, \ldots, e_n\} = \mathrm{span}\{v_1, \ldots, v_n\}$. That completes the proof. $\qquad\square$

**Corollary 3.5.2.** *Every separable inner product space has a countable complete orthonormal system. Thus, by Theorem 3.3.14, every separable Hilbert space has a countable orthonormal basis.*

## 3.6 Additional exercises

**Exercise 3.6.1.** Explain why the following statement is false, and find a meaningful way to fix it: *A subset $S$ of a vector space $V$ is convex if and only if $\frac{1}{2}x + \frac{1}{2}y \in S$ for all $x, y \in S$.*

**Exercise 3.6.2.** Does Theorem 3.2.3 hold when $S$ is not closed? What about Theorem 3.2.5?

**Exercise 3.6.3.** Give an explicit description of the best approximation projection onto the following subsets:

1. $S = \{h \in H : \|h\| \leq 1\} \subset H$.

2. $S = \{(a_n)_{n=0}^\infty : a_n \in [0, \infty)\} \subset \ell^2$.

**Exercise 3.6.4.** Solve the following minimization problems:

1. $\inf_{a,b,c \in \mathbb{R}} \int_{-\pi}^{\pi} |x + a + b \sin x + c \cos x|^2 \, dx$.

2. $\inf_{a,b,c \in \mathbb{R}} \left( \int_{-1}^{1} |e^x + a + bx + cx^2|^2 \, dx + \int_{-\pi}^{\pi} |e^x + b + 2cx|^2 \, dx \right)$.

**Exercise 3.6.5.** Let $H$ be a Hilbert space, $M$ a closed subspace, and $h \in H$. Prove that

$$\inf \{\|h - g\| : g \in M\} = \sup \{\mathrm{Re}\langle h, f \rangle : f \in M^\perp, \|f\| \leq 1\}.$$

We know that the infimum on the left-hand side is actually a minimum, and is attained at a unique point $g \in M$. What about the supremum on the right-hand side?

**Exercise 3.6.6.** Let $A$ be an $m \times n$ matrix over $\mathbb{C}$ and $b \in \mathbb{C}^m$.

1. Suppose that $m > n$ and that $\mathrm{rank}(A) = n$. Typically, the equation $Ax = b$ has no solution. Prove that there exists a unique vector $x \in \mathbb{C}^n$ such that $\|Ax - b\| \leq \|Ay - b\|$ for all $y \in \mathbb{C}^n$, that is, $x$ is the best approximation of a solution that one can get.

2. Show that the unique $x$ is given by the formula $x = (A^*A)^{-1}A^*b$.

3. Suppose that $m < n$ and that $\mathrm{rank}(A) = m$. The equation $Ax = b$ has many solutions. Prove that there exists a unique solution $x \in \mathbb{C}^n$ of minimal norm, and find a formula for $x$ in terms of $A$ and $b$.

(One might approach this exercise using multivariate calculus, but that is not what we are looking for.)

**Exercise 3.6.7.** Let $\{a_n\}_{n \in \mathbb{N}} \subset \mathbb{C}$. Find a meaningful necessary and sufficient condition for the series $\sum_{n \in \mathbb{N}} a_n$ to converge in the sense of Definition 3.3.4.

**Exercise 3.6.8.** Repeat Exercise 2.5.10 for a family of Hilbert spaces $\{H_i\}_{i \in I}$, where $I$ is a not necessarily countable index set.

**Exercise 3.6.9.** Let $G$ be an inner product space and let $\{e_i\}_{i \in I}$ be a complete orthonormal system in $G$.

1. Is $\{e_i\}_{i \in I}$ necessarily a closed system (i.e., does it satisfy the conclusion of Theorem 3.3.14)?

2. If $H$ is the completion of $G$, is $\{e_i\}_{i \in I}$ necessarily a basis for $H$?

**Exercise 3.6.10.** Let $\{a_i\}_{i \in I}$ be such that $\sum_{i \in I} |a_i|^2 < \infty$, and let $\{e_i\}_{i \in I}$ be an orthonormal system in a Hilbert space $H$. Prove that for all $k \in I$,

$$\left\langle \sum_{i \in I} a_i e_i, e_k \right\rangle = a_k.$$

**Exercise 3.6.11** (Legendre polynomials). Applying the Gram-Schmidt process to the sequence $\{1, x, x^2, x^3, \ldots, \}$ in the Hilbert space $L^2[-1,1]$, one obtains an orthonormal system $\{u_0, u_1, u_2, \ldots\}$ consisting of polynomials, called the *normalized Legendre polynomials*.

1. Is this system an orthonormal basis for $L^2[-1,1]$?

2. Carry out one of the following tasks:

    (a) Calculate the first five Legendre polynomials.

    (b) Write a code that computes the first $n$ Legendre polynomials. Plot the graphs of the first five.

    (c) Prove that for every $n \geq 1$,

$$u_n(x) = \sqrt{\frac{2n+1}{2}} \frac{1}{2^n n!} \frac{d^n}{dx^n} \left(x^2 - 1\right)^n.$$

**Exercise 3.6.12.** Let $M$ and $N$ be closed subspaces in a Hilbert space $H$. We define the **sum** of $M$ and $N$ to be the subspace

$$M + N = \{m + n : m \in M, n \in N\}.$$

1. Prove that if $M \perp N$, then $M + N$ is closed.

2. Prove that if $N$ is finite dimensional, then $M + N$ is closed.

3. Can you find another meaningful condition that guarantees closedness of the sum?

4. Give an example of two closed subspaces with $M \cap N = \{0\}$ such that $M + N$ is not closed.

5. Show that

$$(M + N)^\perp = M^\perp \cap N^\perp \quad \text{and} \quad (M \cap N)^\perp = \overline{M^\perp + N^\perp}.$$

**Exercise 3.6.13.** Let $M$ and $N$ be two closed subspaces of a Hilbert space $H$, and let $P = P_M$ and $Q = P_N$ be the orthogonal projections on $M$ and $N$, respectively.

1. Prove that $PQ$ is an orthogonal projection if and only if $P$ and $Q$ commute, i.e., if and only if $PQ = QP$.

2. Prove that if $P$ and $Q$ commute, then $PQ$ is the orthogonal projection onto $M \cap N$, and that $P + Q - PQ$ is the orthogonal projection onto $M + N$.

3. Prove that if $P$ and $Q$ commute, then $M + N$ is closed.

4. Can you give a geometric interpretation of the algebraic condition that $P$ and $Q$ commute?

**Exercise 3.6.14** (The Haar system). For every $n = 1, 2, \ldots$ and $k \in \{0, , \ldots, 2^{n-1} - 1\}$, define functions

$$
\varphi_{k,n}(x) = \begin{cases} 2^{(n-1)/2} & , x \in \left[ \frac{2k}{2^n}, \frac{2k+1}{2^n} \right), \\ -2^{(n-1)/2} & , x \in \left[ \frac{2k+1}{2^n}, \frac{2k+2}{2^n} \right), \\ 0 & , \text{else} . \end{cases}
$$

Let $\varphi_{0,0}$ be the constant function 1 on $[0, 1]$. Prove that the system $\{\varphi_{k,n} : n \in \mathbb{N}, 0 \le k \le 2^n - 1\}$ is a complete orthonormal system in $L^2[0, 1]$. (**Hint:** let $M_N$ denote the finite dimensional subspace of $L^2[0, 1]$ spanned by the characteristic functions of the interval $\left[ \frac{k}{2^N}, \frac{k+1}{2^N} \right)$ (i.e., the step function having value 1 on $\left[ \frac{k}{2^N}, \frac{k+1}{2^N} \right)$ and zero elsewhere) for some $N \in \mathbb{N}$ and all $k \in \{0, \ldots, 2^N - 1\}$. Show that $M_N = \text{span}\{\varphi_{k,n} : 0 \le n \le N\}$, and prove that for every $f \in C([0, 1])$, $P_{M_N} f$ converges to $f$ uniformly as $n \to \infty$).

**Exercise 3.6.15.** Let $\{u_n\}_{n=1}^\infty$ and $\{v_n\}_{n=1}^\infty$ be two orthonormal bases for $\ell^2(\mathbb{N})$. For every $n$, we write $u_n = (u_n(k))_{k=0}^\infty$ and likewise for $v_n$. Define a doubly indexed family of doubly indexed sequences $\{w_{m,n}\}_{m,n=1}^\infty$ by

$$
w_{m,n}(k, l) = u_m(k) v_n(l).
$$

1. Prove that $u_{m,n} \in \ell^2(\mathbb{N} \times \mathbb{N})$ for all $m, n$.

2. Prove that $\{w_{m,n}\}_{m,n=1}^\infty$ is an orthonormal basis for $\ell^2(\mathbb{N} \times \mathbb{N})$.

**Exercise 3.6.16** (Tricky). Let $\{f_n\}_{n=1}^\infty$ and $\{g_n\}_{n=1}^\infty$ be two orthonormal bases for $L^2[0, 1]$ consisting of piecewise continuous functions.

1. Prove that $\{\overline{f_n}\}_{n=1}^\infty$ is an orthonormal basis for $L^2[0, 1]$.

2. Prove that the system of functions $\{h_{m,n}\}_{m,n=1}^\infty$ given by

$$
h_{m,n}(x, y) = f_m(x) g_n(y) \quad , \quad x, y \in [0, 1],
$$

is an orthonormal basis for $L^2([0, 1] \times [0, 1])$.

(**Hint:** it suffices to show that Parseval's identity for all continuous functions.)

The assumption in the above exercise, that the functions $f_n$ and $g_n$ are piecewise continuous, is redundant. The result holds true for any bases of $L^2[0, 1]$, but it is very hard to prove it without measure theory (for example, it is not even clear what $f_m(x) g_n(y)$ means).

# Chapter 4

## Fourier series

### 4.1 Fourier series in $L^2$

Consider the cube $K = [0, 1]^k \subset \mathbb{R}^k$. Let $f$ be an integrable function defined on $K$. For every $n \in \mathbb{Z}^k$, the $n$th **Fourier coefficient of** $f$ is defined to be

$$\hat{f}(n) = \int_K f(x)e^{-2\pi i n \cdot x}dx,$$

where for $n = (n_1, \ldots, n_k)$ and $x = (x_1, \ldots, x_k) \in K$ we write $n \cdot x = n_1 x_1 + \ldots + n_k x_k$. The series

$$\sum_{n \in \mathbb{Z}^k} \hat{f}(n)e^{2\pi i n \cdot x}$$

is called **the Fourier series of** $f$. The basic problem in Fourier analysis is whether one can reconstruct $f$ from its Fourier coefficients, and in particular, under what conditions, and in what sense, does the Fourier series of $f$ converge to $f$.

We are ready to start applying the structure theory of Hilbert spaces that we developed in the previous chapter, together with the Stone-Weierstrass theorem we proved in the introduction, to obtain some results on Fourier series.

### 4.1.1 An approximation result

Recall that we defined $L^2(K)$ to be the completion of the inner product space $C(K)$ with respect to the inner product

$$\langle f, g \rangle = \int_K f(x)\overline{g(x)}dx.$$

This ends up being the same space $L^2(K)$ as one encounters in a course on Lebesgue measure and integration theory. The reader who knows Lebesgue theory may choose either definition for the space $L^2(K)$. The only facts that we will require are, first, that $C(K)$ is dense in $L^2(K)$, and second, that $L^2(K)$ is complete (so it is a Hilbert space).

A simple computation shows that the collection of functions

$$\left\{ e^{2\pi i n \cdot x} \right\}_{n \in \mathbb{Z}^k}$$

is an orthonormal system in $L^2(K)$. We clearly have $\hat{f}(n) = \langle f, e^{2\pi i n \cdot x} \rangle$, i.e., the Fourier coefficients of $f$ are what we defined in the previous chapter to be the generalized Fourier coefficients of $f$ with respect to the system $\{e^{2\pi i n \cdot x}\}_{n \in \mathbb{Z}^k}$.

We let $P$ denote space of all complex trigonometric polynomials, that is, all the finite sums of the form

$$\sum_n a_n e^{2\pi i n \cdot x}.$$

We also let $C_{per}(K)$ denote the set of continuous periodic functions on $K$, that is, the functions $f$ for which

$$f(0, x_2, \ldots, x_k) = f(1, x_2, \ldots, x_k),$$

$$f(x_1, 0, x_3, \ldots, x_k) = f(x_1, 1, x_3, \ldots, x_k),$$

etc., for all $x = (x_1, \ldots, x_k) \in K$ . The spaces $P$ and $C_{per}(K)$ are contained in $C(K)$ and therefore also in $L^2(K)$. We let $\| \cdot \|_\infty$ denote the sup norm in $C(K)$, and we let $\| \cdot \|_2$ denote the Hilbert space norm in $L^2(K)$.

**Lemma 4.1.1.** *$P$ is dense in $C_{per}(K)$ in the $\| \cdot \|_\infty$ norm.*

*Proof.* The case $k = 1$ follows from the trigonometric approximation theorem (Theorem 1.2.3) from the introduction, together with the identities $2i \sin t = e^{it} - e^{-it}$ and $2 \cos t = e^{it} + e^{-it}$. If $k > 1$, then we invoke the multivariate version of this theorem given in Exercise 1.6.7.

Alternatively, we may argue as in the proof of the trigonometric approximation theorem given in Section 1.3, and identify $C_{per}(K)$ with $C(\mathbb{T}^k)$ via the map

$$\Phi : C(\mathbb{T}^k) \to C_{per}(K) \quad , \quad \Phi(f)(x) = f\left(e^{2\pi i x_1}, \ldots, e^{2\pi i x_k}\right).$$

Here $\mathbb{T} = \{z \in \mathbb{C} \mid |z| = 1\}$ and $\mathbb{T}^k = \mathbb{T} \times \cdots \times \mathbb{T}$ ($k$ times). We may then consider $P$ as a self-adjoint subalgebra of $C(\mathbb{T}^k)$ containing the constants, and it is easy to verify that $P$ separates points, so we may apply the complex version of the Stone-Weierstrass theorem (Theorem 1.4.5) to conclude that the closure of $P$ in the $\| \cdot \|_\infty$ norm is equal to $C(\mathbb{T}^k) \cong C_{per}(K)$. $\square$

**Lemma 4.1.2.** *$P$ is dense in $C_{per}(K)$ in the $\| \cdot \|_2$ norm.*

*Proof.* Clearly, for every $f \in C(K)$ we have

$$\|f\|_2 = \left(\int_K |f(x)|^2 dx\right)^{1/2} \leq \left(\int_K \|f\|_\infty^2 dx\right)^{1/2} = \|f\|_\infty,$$

and the result follows from Lemma 4.1.1. Indeed, if $f \in C_{per}(K)$ is given and $p \in P$ satisfies $\|p - f\|_\infty < \epsilon$, then

$$\|p - f\|_2 \leq \|p - f\|_\infty < \epsilon.$$

$\square$

**Lemma 4.1.3.** $C_{per}(K)$ *is dense in* $C(K)$ *in the* $\|\cdot\|_2$ *norm.*

*Proof.* Let $\epsilon > 0$, and set $L = [\epsilon, 1 - \epsilon]^k$. Let $g \in C(K)$ be a function that satisfies:

1. $0 \le g \le 1$,

2. $g\big|_L = 1$,

3. $g\big|_{\partial K} = 0$.

Such a function is easy to construct explicitly: for example

$$g(x) = \epsilon^{-1} \left( d(x, \partial K) \wedge \epsilon \right),$$

where $d(x, \partial K) = \inf\{\|x - y\| : y \in \partial K\}$. If $f \in C(K)$, then $fg \in C_{per}(K)$ and

$$\|f - fg\|_2^2 = \int_K |f|^2 |1 - g|^2 dx \le \|f\|_\infty^2 \int_{K \setminus L} 1 dx$$

and the right-hand side is less than $\|f\|_\infty^2 (1 - (1 - 2\epsilon)^k)$, which can be made as small as we wish. $\qquad\square$

**Corollary 4.1.4.** $P$ *is dense in* $L^2(K)$.

*Proof.* The corollary is immediate from Lemmas 4.1.2 and 4.1.3, combined with the fact that $C(K)$ is dense in $L^2(K)$. $\qquad\square$

## 4.1.2 Norm convergence of Fourier series

In the following theorem, we will write $|n| = |n_1| + \ldots + |n_k|$, for all $n \in \mathbb{Z}^k$.

**Theorem 4.1.5** ($L^2$ convergence of Fourier series). *For any* $f \in L^2(K)$, *the Fourier series of* $f$ *converges to* $f$ *in the* $\|\cdot\|_2$ *norm. In particular,* $\{e^{2\pi i n \cdot x}\}_{n \in \mathbb{Z}^k}$ *is a complete orthonormal system, and the following hold:*

1. $\|f\|_2^2 = \sum |\hat{f}(n)|^2$,

2. $\lim_{N \to \infty} \|f(x) - \sum_{|n| \le N} \hat{f}(n) e^{2\pi i n \cdot x}\|_2 = 0$.

*Proof.* The theory is so neat and tight that one can give several slightly different quick proofs.

First proof: By Corollary 4.1.4, we have that (3) of Proposition 3.3.11 holds. Therefore, the equivalent conditions (1) and (2) of that Proposition hold, which correspond to assertions (1) and (2) in the theorem above. Completeness is immediate from either (1) or (2).

Second proof: By Corollary 4.1.4, the system $\{e^{2\pi i n \cdot x}\}_{n \in \mathbb{Z}^k}$ is complete. Indeed, assume that $f \perp \{e^{2\pi i n \cdot x}\}_{n \in \mathbb{Z}^k}$. Then $f \perp P$. The corollary implies that there is a sequence in $P \ni p_i \to f$. Thus

$$\langle f, f \rangle = \lim_{i \to \infty} \langle f, p_i \rangle = 0,$$

whence $f = 0$. Now Theorem 3.3.14 gives the result. $\qquad\square$

**Remark 4.1.6.** Theorem 4.1.5 gives us another way of thinking of the space $L^2[0,1]$: it is the space of Fourier series $\sum_n a_n e^{2\pi i t}$ where $(a_n)_{n \in \mathbb{Z}} \in \ell^2(\mathbb{Z})$ (the space $\ell^2(\mathbb{Z})$ was defined in Example 2.2.20).

Theorem 4.1.5, although interesting, elegant and useful, leaves a lot of questions unanswered. For example, what about pointwise convergence? For functions in $L^2[a,b]$, only almost everywhere convergence makes sense, and it is a fact, proved by Carleson [5], that the Fourier series of every $f \in L^2[a,b]$ converges *almost everywhere to* $f$. Carleson's theorem requires far more delicate analysis than the norm convergence result that we obtained, and is beyond the scope of this course. Another natural question is: what about uniform convergence? It turns out that Theorem 4.1.5 is powerful enough to imply some neat results.

From now on, until the end of the chapter, we will concentrate on functions and spaces of functions in one variable.

Let $C^1([0,1])$ denote the space of all continuously differentiable functions, that is, the space of all functions $f : [0,1] \to \mathbb{C}$ that are differentiable on $(0,1)$ and one sided differentiable at the endpoints, such that the derivative $f'$ is continuous on $[0,1]$.

**Theorem 4.1.7.** *For every* $f \in C_{per}([0,1]) \cap C^1([0,1])$, *the Fourier series of* $f$ *converges uniformly to* $f$:

$$\lim_{N \to \infty} \sup_{t \in [0,1]} \left| \sum_{n=-N}^{N} \hat{f}(n) e^{2\pi i n t} - f(t) \right| = 0.$$

*Proof.* Consider the Fourier coefficients $\widehat{f'}(n)$ of $f'$. By integration by parts, we find that

$$\widehat{f'}(n) = 2\pi i n \hat{f}(n),$$

for all $n \in \mathbb{Z}$. Since $f' \in L^2[0,1]$, we have by Bessel's inequality $\sum |\widehat{f'}(n)|^2 < \infty$, so by Cauchy-Schwarz

$$\sum_{n \neq 0} |\hat{f}(n)| \leq \left( \sum_{n \neq 0} |\widehat{f'}(n)|^2 \right)^{1/2} \left( \sum_{n \neq 0} \frac{1}{4\pi^2 n^2} \right)^{1/2} < \infty.$$

By the Weierstrass M-test, the sequence of partial sums $\sum_{|n| \leq N} \hat{f}(n) e^{2\pi i n t}$ converges uniformly, as $N \to \infty$, to some continuous function $g$. It follows that the sequence of partial sums converges to $g$ in the $L^2$ norm as well. But by Theorem 4.1.5, the partial sums $\sum_{|n| \leq N} \hat{f}(n) e^{2\pi i n t}$ converge to $f$ in the norm of $L^2[0,1]$. Thus $f = g$ and the proof is complete. $\square$

In fact, the conditions of Theorem 4.1.7 can be weakened somewhat.

**Definition 4.1.8.** We let $PC^1[a,b]$ denote the space of all functions $f : [a,b] \to \mathbb{C}$ such that

1. $f \in PC[a, b]$,

2. $f$ is differentiable at all but finitely many points in $[a, b]$,

3. The derivative $f'$ (which is defined at all but finitely many points) is also in $PC[a, b]$.

**Exercise 4.1.9.** Check that the proof of Theorem 4.1.7 also works for $f \in C_{per}([0, 1]) \cap PC^1[0, 1]$. For this, prove that the formula for integration by parts

$$\int_a^b f'(t)g(t)dt = f(b)g(b) - f(a)g(a) - \int_a^b f(t)g'(t)dt,$$

holds whenever $f \in C([a, b]) \cap PC^1[a, b]$ and $g \in C^1([a, b])$. (**Hint:** simply break the interval into parts where $f$ is continuously differentiable.)

**Remark 4.1.10.** Note that if a Fourier series converges uniformly, then the limit must be in $C_{per}(K)$. On the other hand, there are functions in $C_{per}([0, 1])$ whose Fourier series diverges on a dense set of points in $[0, 1]$ (see [35, Chapter VIII]), so, in particular, the partial sums $\sum_{|n| \leq N} \hat{f}(n)e^{2\pi int}$ of the Fourier series of a general continuous and periodic function need not approximate it uniformly. However, we will see below that continuous and periodic functions *can* be approximated uniformly, by summing the Fourier series in a clever way.

### 4.1.3 An example

**Example 4.1.11.** Let $g \in L^2[0, 1]$ be given by $g(t) = t$. We find by a direct calculation that the Fourier coefficients of $g$ are given by

$$\hat{g}(0) = \frac{1}{2},$$

and

$$\hat{g}(n) = \frac{i}{2\pi n}, \quad n \neq 0.$$

Thus

$$t = \frac{1}{2} + \sum_{n \in \mathbb{Z} \setminus \{0\}} \frac{i}{2\pi n} e^{2\pi int} = \frac{1}{2} - \sum_{n=1}^{\infty} \frac{\sin(2\pi nt)}{\pi n}, \tag{4.1}$$

where equality means that both sides represent the same element in $L^2[0, 1]$.

By Parseval's identity,

$$\int_0^1 |g(t)|^2 dx = \sum_{n \in \mathbb{Z}} |\hat{g}(n)|^2.$$

The left-hand side is $1/3$, and the right-hand side is $1/4 + 2\sum_{n=1}^{\infty} \frac{1}{4\pi^2 n^2}$, so, after rearranging, we find that

$$\sum_{n=1}^{\infty} \frac{1}{n^2} = \frac{\pi^2}{6}.$$

What about pointwise convergence? By the exercise below, the Fourier series of $t$ converges (at least when summed in an appropriate order) for every $t \in [0, 1]$. But what is the limit? Can we say that the right-hand side of (4.1) converges to $t$ for every $t \in [0, 1]$? We will return to this question below.

**Exercise 4.1.12.** Prove that the limit

$$\lim_{N \to \infty} \sum_{0 < |n| < N} \frac{1}{2\pi i n} e^{2\pi i n t}$$

exists (and is finite) for every $t \in [0, 1]$. (**Hint:** You need not calculate the limit for $t \in (0, 1)$, you can use Dirichlet's test for convergence of series. For $t = 0, 1$, keep in mind that in the partial sums $n$ goes from $-N$ to $N$.)

---

## 4.2    Pointwise convergence of Fourier series (Dirichlet's theorem)

### 4.2.1    Fourier series on different intervals

In the previous section we discussed the convergence of the Fourier series of a function $f \in L^2[0, 1]$ in the $L^2$ norm (Theorem 4.1.5), as well as the uniform convergence of the Fourier series of a function $f \in C_{per}([0, 1]) \cap C^1([0, 1])$ (Theorem 4.1.7). It should be clear that there is nothing special about the interval $[0, 1]$. If we wish to work with an interval $[a, b]$ instead, we can consider the space $L^2[a, b]$ with the inner product

$$\langle f, g \rangle = \frac{1}{b - a} \int_a^b f(t) \bar{g}(t) dt.$$

An orthonormal basis for this space is given by $\left\{ e^{i \frac{2\pi n}{b-a} t} \right\}_{n \in \mathbb{Z}}$. For a function $f \in L^2[a, b]$, we define the **Fourier coefficients**

$$\hat{f}(n) = \frac{1}{b - a} \int_a^b f(t) e^{-i \frac{2\pi n}{b-a} t}(t) dt.$$

The series $\sum_{n \in \mathbb{Z}} \hat{f}(n) e^{i \frac{2\pi n}{b-a} t}$ is called **the Fourier series of $f$**. It is straightforward that whatever convergence theorems we prove for Fourier series in one interval, we get analogous theorems in all other intervals.

**Exercise 4.2.1.** Prove that the appropriate versions of Theorems 4.1.5 and 4.1.7 hold true for Fourier series in any interval.

## 4.2.2 The Dirichlet kernel and Dirichlet's theorem

The next result we discuss is Dirichlet's theorem — a theorem on the pointwise convergence of Fourier series. It will be convenient for us to consider functions on the interval $[-\pi, \pi]$, and we normalize the inner product in $L^2[-\pi, \pi]$ to be

$$\langle f, g \rangle = \frac{1}{2\pi} \int_{-\pi}^{\pi} f(t)\bar{g}(t)dt.$$

The orthonormal basis with respect to which Fourier series are constructed is then $\{e^{int}\}_{n\in\mathbb{Z}}$, and the Fourier coefficients of a function $f \in L^2[-\pi, \pi]$ are given by

$$\hat{f}(n) = \frac{1}{2\pi} \int_{-\pi}^{\pi} f(t)e^{-int}(t)dt.$$

We now define

$$S_N(f)(t) = \sum_{|n|\leq N} \hat{f}(n)e^{int}.$$

The functions $S_N(f)$ are called **the partial sums** of the Fourier series of $f$.

We will identify every $f \in PC[-\pi, \pi]$ with a $2\pi$-periodic function defined on the whole real line. If $f(\pi) = f(-\pi)$, then there is a unique such periodic extension. If $f(\pi) \neq f(-\pi)$, then there is some ambiguity because it is not clear what value we want to give the function at the points $\pi k$ for odd $k$. This ambiguity causes no trouble, and will be washed away when our functions are plugged into integrals.

Changing order of integration and summation, and then applying a change of variables, we have

$$S_N(f)(x) = \sum_{n=-N}^{N} \left( \frac{1}{2\pi} \int_{-\pi}^{\pi} f(t)e^{-int}(t)dt \right) e^{inx}$$

$$= \frac{1}{2\pi} \int_{-\pi}^{\pi} \left( \sum_{n=-N}^{N} e^{in(x-t)} \right) f(t)dt$$

$$= \frac{1}{2\pi} \int_{-\pi}^{\pi} D_N(x-t)f(t)dt$$

$$= \frac{1}{2\pi} \int_{-\pi}^{\pi} D_N(t)f(x-t)dt,$$

where we have introduced the notation $D_N(t) = \sum_{n=-N}^{N} e^{int}$.

The sequence of functions $\{D_N\}_{N=0}^{\infty}$ is called the **Dirichlet kernel**. Straightforward calculations using the formula for the sum of a geometric series shows that

$$D_N(t) = \frac{\sin\left((N+1/2)t\right)}{\sin(t/2)}. \tag{4.2}$$

It is interesting to plot graphs of $D_N$ for several $N$ (try it!).

**Exercise 4.2.2.** Prove the above formula for the Dirichlet kernel $D_N(t)$.

**Definition 4.2.3.** Let $g, h$ be $2\pi$-periodic functions on $\mathbb{R}$, and assume that they are integrable on every finite subinterval of $\mathbb{R}$. The **convolution** of $g$ and $h$ is the ($2\pi$-periodic) function

$$g * h(x) = \frac{1}{2\pi} \int_{-\pi}^{\pi} g(t)h(x - t)dt.$$

With this terminology, we have that $S_N(f)(x) = D_N * f(x)$, that is, the partial sums of the Fourier series of $f$ are given by the convolution of $f$ with the Dirichlet kernel.

The following theorem treats functions in the space $PC^1[-\pi, \pi]$ (recall Definition 4.1.8). We will let

$$f(t_-) = \lim_{0 < h \to 0} f(t - h) \quad \text{and} \quad f(t_+) = \lim_{0 < h \to 0} f(t + h)$$

denote the one-sided limits. These one-sided limits exist at every $t \in (-\pi, \pi)$ for every $f \in PC[-\pi, \pi]$. At the point $-\pi$ the left one-sided limit $f(-\pi_-)$ is defined to be $f(\pi_-)$, and similarly at the point $\pi$ the right one-sided limit $f(\pi_+)$ is defined to be $f(-\pi_+)$.

**Theorem 4.2.4** (Dirichlet's theorem). *Let $f \in PC^1[-\pi, \pi]$. At every $t \in [-\pi, \pi]$ the partial sums of the Fourier series converge, and the limit is*

$$\lim_{N \to \infty} S_N(f)(t) = \frac{f(t_-) + f(t_+)}{2}.$$

The proof of Dirichlet's theorem is deferred to Section 4.4.

**Remark 4.2.5.** In particular, we see that under the assumptions of the theorem, $S_N(f)(t)$ converges to $f(t)$ at every point of continuity (what does this mean for $t = \pm\pi$?).

---

## 4.3   Fejér's theorem

It is a fact (see [35, Chapter VIII]), that there are functions in $C_{per}([0, 1])$ whose Fourier series diverge on a dense set of points in $[0, 1]$. In particular, contrary to the intuition provided by Dirichlet's theorem, the Fourier series of a continuous and periodic function need not converge pointwise to its limit.

If $\{a_n\}$ is a sequence of numbers and $a_n \to L$, then the so-called *Cesàro sums* also converge to $L$:

$$\frac{a_0 + \ldots + a_N}{N + 1} \to L.$$

On the other hand, it may happen that the Cesàro sums converge but the ordinary series does not. Thus, forming the Cesàro sums only improves the chances of convergence. This rather loose heuristic turns out to work in the setting of Fourier series.

Before proceeding, let us stress that we will continue to identify a function $f \in PC[-\pi, \pi]$ with a $2\pi$ periodic function (well-defined except perhaps at points of the form $\pi k$ for odd $k$). For every function $f \in PC[-\pi, \pi]$ and every $x \in [-\pi, \pi]$ we form the **Cesàro sums**

$$\sigma_N(f)(x) = \frac{1}{N+1} \sum_{n=0}^{N} S_n(f)(x),$$

in hope of improving the convergence.

We find that

$$\sigma_N(f)(x) = \frac{1}{N+1} \sum_{n=0}^{N} D_n * f(x)$$
$$= K_N * f(x),$$

where $K_N = \frac{1}{N+1} \sum_{n=0}^{N} D_n$ is called the **Fejér kernel**. Since $D_n(t) = \sum_{n=-N}^{N} e^{int}$, we can compute

$$K_N(t) = \frac{1}{N+1} \sum_{n=0}^{N} D_n(t) = \sum_{n=-N}^{N} \left(1 - \frac{|n|}{N+1}\right) e^{int}. \tag{4.3}$$

Cleverly noting that

$$\frac{1}{N+1} \left| \sum_{n=0}^{N} e^{int} \right|^2 = \sum_{n=-N}^{N} \left(1 - \frac{|n|}{N+1}\right) e^{int},$$

one can see that

$$K_N(t) = \frac{1}{N+1} \frac{\sin^2\left(\frac{(N+1)t}{2}\right)}{\sin^2\left(\frac{t}{2}\right)}. \tag{4.4}$$

**Lemma 4.3.1.** *For all $N$, $K_N$ has the following properties:*

1. *$K_N(t) \geq 0$ for all $t$,*

2. *$K_N$ is periodic with period $2\pi$,*

3. *$\frac{1}{2\pi} \int_{-\pi}^{\pi} K_N(t)dt = 1$,*

4. *For every $\delta > 0$, $K_N$ converges uniformly to $0$ on $[-\pi, -\delta] \cup [\delta, \pi]$.*

*Proof.* Items (1) and (2) follow immediately from (4.4), and (3) follows from (4.3). Since for $\delta > 0$ the function $\sin^2\left(\frac{t}{2}\right)$ is continuous and does not vanish on $[-\pi, -\delta] \cup [\delta, \pi]$, the final assertion also follows from (4.4). □

**Theorem 4.3.2** (Fejér's theorem). *Let $f \in PC[-\pi, \pi]$.*

*1. At every point $x \in [-\pi, \pi]$ where $f$ is continuous, $\sigma_N(f)(x) \xrightarrow{N \to \infty} f(x)$.*

*2. If $f \in C_{per}([-\pi, \pi])$, then the Cesàro sums converge uniformly to $f$.*

*Proof.* Fix $\epsilon > 0$ and a point of continuity $x \in [-\pi, \pi]$. We consider the difference $\sigma_N(f)(x) - f(x) = K_N * f(x) - f(x)$. We rewrite this as

$$\frac{1}{2\pi} \int_{-\pi}^{\pi} K_N(t) f(x - t) dt - \frac{1}{2\pi} \int_{-\pi}^{\pi} K_N(t) f(x) dt$$

$$= \frac{1}{2\pi} \int_{-\pi}^{\pi} K_N(t)(f(x - t) - f(x)) dt.$$

Choose $\delta > 0$ such that $|x - y| < \delta$ implies $|f(x) - f(y)| < \epsilon$. We split the integral into three parts, and estimate

$$|\sigma_N(f)(x) - f(x)| \leq \frac{1}{2\pi} \int_{-\pi}^{\pi} K_N(t) |f(x - t) - f(x)| dt$$

$$= \frac{1}{2\pi} \int_{-\delta}^{\delta} K_N(t) |f(x - t) - f(x)| dt$$

$$+ \frac{1}{2\pi} \int_{-\pi}^{-\delta} K_N(t) |f(x - t) - f(x)| dt$$

$$+ \frac{1}{2\pi} \int_{\delta}^{\pi} K_N(t) |f(x - t) - f(x)| dt.$$

The first integal is less than $\frac{1}{2\pi} \int_{-\delta}^{\delta} K_N(t) \epsilon dt \leq \frac{\epsilon}{2\pi} \int_{-\pi}^{\pi} K_N(t) dt = \epsilon$ by choice of $\delta$. The second and third integrals are less than $\frac{\|f\|_{\infty}}{\pi} \int_{\delta}^{\pi} K_N(t) dt$, and this can be made to be less than $\epsilon$ if one takes $N$ sufficiently large, because the integrand converges uniformly to 0 on the domain of integration. This shows that $\sigma_N(f)(x) \xrightarrow{N \to \infty} f(x)$.

Finally, if $f \in C_{per}([-\pi, \pi])$, then the periodic extension of $f$ to $\mathbb{R}$ is uniformly continuous, and $\delta$ can be chosen independently of $x$. Going over the above proof one sees then that $\sigma_N(f) \xrightarrow{N \to \infty} f$ uniformly. $\square$

**Remark 4.3.3.** Any sequence $\{F_N\}$ with the four properties listed in Lemma 4.3.1 is said to be a **summability kernel**, and one can show, along the lines of the proof above, that for every summability kernel $\{F_N\}$, and for any continuous and periodic function $f$, the sequence $F_N * f$ converges uniformly to $f$ (see also Exercise 4.5.16).

**Example 4.3.4.** We will now show that the conclusion of Dirichlet's theorem is valid for the function $g(t) = t$ on $[0, 1]$ treated in Example 4.1.11. For $t = 0$ and $t = 1$ we can see directly that $S_N(g)(t) \xrightarrow{N \to \infty} \frac{1}{2}$. If $t \neq 0, 1$, then $f$ is continuous at $t$, so by Fejér's theorem $\sigma_N(g)(t)$ converges to $f(t) = t$. But by Exercise 4.1.12, the partial sums of the Fourier series $S_N(g)(t)$ converge, so they must converge to the same value as the Cesàro sums $\sigma_N(f)(t)$. Thus $S_N(g)(t) \xrightarrow{N \to \infty} t = f(t)$ for all $t \neq 0, 1$.

## 4.4 *Proof of Dirichlet's theorem

We will now prove Theorem 4.2.4, using nothing but the results from Exercise 4.1.9, together with the fact, just proved in Example 4.3.4, that Dirichlet's theorem holds for the function $g(t) = t$. Note that we stated Dirichlet's theorem in the interval $[-\pi, \pi]$, but we are going to prove it in the interval $[0, 1]$ (recall the comments in Subsection 4.2.1).

Let $f \in PC^1[0,1]$. The proof is by induction on the number $n$ of points of discontinuity of $f$ (including the "point" $0 = 1$). If $n = 0$, then $f \in C_{per}([0,1]) \cap PC^1[0,1]$, and we are in the situation treated in Exercise 4.1.9; thus $S_N(f)$ converges uniformly to $f$.

Suppose now that the number of points of discontinuity of $f$ is $n \geq 1$, and that Dirichlet's theorem holds for all functions in $PC^1[a,b]$ which have $n-1$ or fewer points of discontinuity. Without loss of generality, we may assume that $f(0) = f(0_+)$ and that $f(0_+) \neq f(1_-)$ accounts for one point of discontinuity (why?). Further, by adding a constant to $f$ and multiplying by a nonzero constant, we may assume that $f(0_+) = 0$ and that $f(1_-) = 1$. Now we define a new function $h = f - g$, where $g(t) = t$. The function $h$ is in $PC^1[0,1]$, and has only $n - 1$ points of discontinuity. Since

$$S_N(f)(t) = S_N(h)(t) + S_N(g)(t)$$

for all $t$, the result for $f$ follows readily by the inductive hypothesis. That concludes the proof of Dirichlet's theorem.

## 4.5 Additional exercises

**Exercise 4.5.1.** Let $f \in L^2(K)$ (as above, $K = [0,1]^k$), and suppose that the Fourier series $\sum_{n \in \mathbb{Z}^k} \hat{f}(n) e^{2\pi i n \cdot x}$ associated with $f$ converges uniformly. Prove that $f \in C_{per}(K)$.

**Exercise 4.5.2.** Establish the identities (4.2) and (4.4).

**Exercise 4.5.3.** Fill in the missing details in the proof of Dirichlet's theorem given in Section 4.4.

**Exercise 4.5.4** ("Real" trigonometric series). When dealing with real-valued functions, it may be more convenient to use an orthonormal system consisting of real-valued functions. The classical example is the system consisting of cosines and sines. For simplicity, we consider the space $L^2[-\pi, \pi]$ with the inner product $\langle f, g \rangle = \frac{1}{\pi} \int_{-\pi}^{\pi} f(t)\bar{g}(t)dt$ (it is clear that the inner product without the normalization constant $\frac{1}{\pi}$ gives rise to the same Hilbert space, just with an inner product that differs by a constant).

1. Prove that $\left\{ \frac{1}{\sqrt{2}}, \cos x, \sin x, \cos 2x, \sin 2x, \ldots \right\}$ is a complete orthonormal system in $L^2[-\pi, \pi]$.

2. Deduce that for every $f \in L^2[-\pi, \pi]$, there is a Fourier series representation

$$ f(x) = \frac{a_0}{2} + \sum_{n=1}^{\infty} a_n \cos nx + b_n \sin nx, \qquad (4.5) $$

where the series converges in the $L^2$ norm (the variable $x$ appearing in the formula is there just so we can write the functions $\sin nx$ and $\cos nx$). Find explicit formulas for $a_n$ $b_n$, and explain also why

$$ \frac{1}{\pi} \int_{-\pi}^{\pi} |f(t)|^2 dt = \frac{|a_0|^2}{2} + \sum_{n=1}^{\infty} |a_n|^2 + |b_n|^2. $$

3. What can you say about pointwise or uniform convergence?

4. Show that if $f$ is a piecewise continuous function which is even (meaning $f(x) = f(-x)$ for all $x \in [-\pi, \pi]$), then $b_n = 0$ for all $n$, and that

$$ a_n = \frac{2}{\pi} \int_0^{\pi} f(t) \cos nt\, dt \qquad \text{for all } n = 0, 1, 2, \ldots. \qquad (4.6) $$

(Here we use the convention that for $n = 0$, the function $\cos nx$ is understood as the constant function $1 = \cos 0$.)

5. Likewise, show that if $f \in PC[-\pi, \pi]$ is odd (meaning $f(x) = -f(-x)$ for all $x \in [-\pi, \pi]$) then $a_n = 0$ for all $n$. Show that if $f$ is continuous and periodic, then the converse is also true. What happens if $f$ is not assumed periodic?

The series in (4.5) is sometimes called the **real** Fourier series, whereas the series $\sum_{n \in \mathbb{Z}} \hat{f}(n)e^{inx}$ is called the **complex** Fourier series. Real and complex Fourier series can be developed for functions on any interval. Find the appropriate formulas for their representation.

**Exercise 4.5.5.** Let $f(x) = \cos(x/2)$ on $[-\pi, \pi]$.

1. Compute the real trigonometric series of $f$ on $[-\pi, \pi]$.

2. Calculate the limit

$$\lim_{N\to\infty} \sum_{n=1}^{N} \frac{1}{(n^2 - \frac{1}{4})^2}.$$

3. Calculate the limit

$$\lim_{N\to\infty} \frac{1}{N+1} \left[ \frac{2}{\pi} + \sum_{m=1}^{N} \left( \frac{2}{\pi} + \sum_{n=1}^{m} \frac{-1}{\pi(n^2 - \frac{1}{4})} \right) \right].$$

**Exercise 4.5.6** (Sine and cosine series). Fourier series are not the only useful trigonometric series representing functions. Consider $L^2[0, \pi]$ with the inner product $\langle f, g \rangle = \frac{2}{\pi} \int_0^\pi f(t)\bar{g}(t)dt$.

1. Prove that $\left\{ \frac{1}{\sqrt{2}}, \cos x, \cos 2x, \cos 3x, \ldots \right\}$ is a complete orthonormal system in $L^2[0, \pi]$, and that the series

$$\sum_{n=0}^{\infty} a_n \cos nx$$

   converges in the norm to $f$ for every $f \in L^2[0, \pi]$, where $a_n$ is given by (4.6). This series is called the **cosine series** of $f$.

2. Prove that $\{\sin x, \sin 2x, \sin 3x, \ldots\}$ is a complete orthonormal system in $L^2[0, \pi]$. The corresponding series is called the **sine series** for a given $L^2$ function.

3. Prove that if $f \in C^1[0, \pi]$, then the cosine series of $f$ converges uniformly to $f$. What about the sine series?

**Exercise 4.5.7.** Let $\{u_n\}_{n=1}^\infty$ be an orthonormal system of functions in $L^2[0, 1]$. Prove that $\{u_n\}_{n=1}^\infty$ is complete if and only if for every $k = 1, 2, \ldots$,

$$\sum_{n=1}^{\infty} \left| \int_0^1 \sin(k\pi t)u_n(t)dt \right|^2 = \frac{1}{2}.$$

(If you are uneasy about writing down that integral for $L^2$ functions, you can assume that the functions $u_n$ are all piecewise continuous.)

**Exercise 4.5.8** (Reem[1]). Let $f : \mathbb{R} \to \mathbb{C}$ be a periodic function with period $2\pi$, such that $f|_{[a,b]}$ is piecewise continuous for every interval $[a, b] \subset \mathbb{R}$. Prove that if

$$f(x + y) = f(x)f(y) \quad \text{for all } x, y \in \mathbb{R},$$

then either $f \equiv 0$, or there exists $k \in \mathbb{Z}$ such that $f(x) = e^{ikx}$.

---

[1]This classical exercise was suggested to me by Daniel Reem. See [25] for a quick proof of this result in greater generality.

**Exercise 4.5.9.** The purpose of this exercise is to explore the connection between the degree of smoothness of a function and the decay rate of its Fourier coefficients. Typically, a very smooth function has Fourier coefficients that decay fast, and functions with Fourier coefficients that decay fast tend to be smooth.

1. Prove that if $f \in PC[0,1]$, and for some $k \in \mathbb{N}$, then the condition $\sum n^k |\hat{f}(n)| < \infty$ implies that $f$ is equivalent in $PC[0,1]$ to a function $g$ has $k$ continuous and periodic derivatives (recall that this means that there is some $g \in C^k([0,1])$, such that $f(x) = g(x)$ foll $x \in [0,1]$ except finitely many exceptions).

2. Prove a result in the converse direction.

**Exercise 4.5.10** (Rudin[2]). Fix $\Delta \in (0,\pi)$. Let $f : [0,\pi] \to \mathbb{C}$ be defined by

$$f(x) = \begin{cases} 1, & x \in [0,\Delta] \\ 0, & x \in (\Delta,\pi] \end{cases}.$$

1. Compute the cosine series of $f$.

2. Compute the sum of the series

$$\sum_{n=1}^{\infty} \frac{\sin n\Delta}{n} \quad \text{and} \quad \sum_{n=1}^{\infty} \left(\frac{\sin n\Delta}{n}\right)^2.$$

3. Use the above to obtain the values of the following integrals

$$\int_0^{\infty} \frac{\sin x}{x} dx = \frac{\pi}{2} \quad \text{and} \quad \int_0^{\infty} \left(\frac{\sin x}{x}\right)^2 dx = \frac{\pi}{2}.$$

**Exercise 4.5.11.** For any $s > 1$, define $\zeta(s) = \sum_{n=1}^{\infty} n^{-s}$. In Example 4.1.11 we calculated

$$\zeta(2) = \sum_{n=1}^{\infty} \frac{1}{n^2} = \frac{\pi^2}{6}.$$

In fact, Fourier series can used to calculate the value of $\zeta(2k) = \sum_{n=1}^{\infty} \frac{1}{n^{2k}}$ for all $k = 1, 2, 3, \ldots$. Do either one of the following tasks:

1. Find the value of $\zeta(4)$ and of $\zeta(6)$.

2. Prove that for any $k = 1, 2, \ldots$, there is a rational number $R_k$ such that

$$\zeta(2k) = R_k \pi^{2k}.$$

---

[2]This is taken from Exercise 12 on page 198 [28].

(**Hint:** calculate the real Fourier series of the functions $x^{2k}$ on the interval $[-\pi, \pi]$. Express the Fourier coefficients of $x^{2k}$ in terms of those of $x^{2(k-1)}$.)

**Exercise 4.5.12.** We define the convolution of two functions $f$ and $g$ in $PC[0, 1]$ by

$$f * g(x) = \int_0^1 f(x - t)g(t)dt,$$

where we extend both $f$ and $g$ to periodic functions with period 1 on the real line, in order that the integral make sense.

1. Prove that if $f, g \in PC[0, 1]$, then

$$\widehat{f * g}(n) = \hat{f}(n)\hat{g}(n), \quad n \in \mathbb{Z}.$$

2. Try to make sense of the following: If $f$ and $g$ are sufficiently nice functions, then

$$\widehat{fg}(n) = \sum_{k \in \mathbb{Z}} \hat{f}(n - k)\hat{g}(k), \quad m \in \mathbb{Z}.$$

(See also Exercise 5.7.12.)

**Exercise 4.5.13** (Gibbs phenomenon). Consider the function $f(t) = t - \lfloor t \rfloor$, which is periodic with period 1 on $\mathbb{R}$ (here, $\lfloor t \rfloor$ denotes the largest integer that is less than or equal to $t$). In Example 4.1.11, we computed the Fourier series of $f$, and we saw that it is given by

$$t = \frac{1}{2} - \sum_{n=1}^{\infty} \frac{\sin(2\pi nt)}{\pi n}.$$

Plot the graphs of $f$, of $S_N(f)$ for $N = 1, 2, 4, 10, 20, 40, 60, 80, 100$, and of $\lim_{N \to \infty} S_N(f)$, over the interval $[-2, 2]$. Note that the graph of $S_N(f)$ becomes a better and better approximation of the graph of $f$, except that near the points of jump discontinuity of $f$, the graph of $S_N(f)$ "overshoots" and then "undershoots". We shall analyze this phenomenon.

1. Show that

$$\lim_{N \to \infty} \sum_{n=1}^{N} \frac{\sin\left(\frac{\pi n}{N}\right)}{n\pi} = \int_0^1 \frac{\sin(\pi x)}{\pi x} dx \approx 0.58949$$

(**Hint:** for the first equality, interpret the sum as a Riemann sum.)

2. Deduce that $S_N(f)\left(\frac{-1}{2N}\right) \xrightarrow{N \to \infty} 1.089....$ This shows that the graphs you plotted did not lie, and that indeed the partial Fourier series have a spike whose height above the graph of the function is about 1.09 the size of the jump. Moreover, as $N$ gets larger, the spike moves closer to the point of discontinuity, but is never smoothed away.

3. Now let $f$ be any function in $PC^1[0,1]$. Let $t \in [0,1]$ be a point of discontinuity of the periodic extension of $f$, and let $J = f(t_+) - f(t_-)$ be the size of the jump at $t$. Prove that

$$S_N(f)\left(t + \frac{1}{2N}\right) \xrightarrow{N \to \infty} f(x_+) + (0.0894...) \times J$$

and

$$S_N(f)\left(t - \frac{1}{2N}\right) \xrightarrow{N \to \infty} f(x_-) - (0.0894...) \times J.$$

This phenomenon is called *Gibbs phenomenon*.

4. Think fast: if you use Fejér sums instead of Dirichlet sums, will that eliminate the overshoot?

**Exercise 4.5.14.** Let $f \in PC[-\pi, \pi]$, and fix $x_0 \in (-\pi, \pi)$. Prove that if $\left| \frac{f(x) - f(x_0)}{x - x_0} \right|$ is bounded in some punctured neighborhood of $x_0$, then $S_N(f)(x_0) \xrightarrow{N \to \infty} f(x_0)$. (**Hint:** write

$$D_N * f(x_0) - f(x_0) = \frac{1}{2\pi} \int_{-\pi}^{\pi} D_N(t)(f(x_0 - t) - f(x_0))dt,$$

and from then on it is just a little dance with epsilons and deltas.)

**Exercise 4.5.15.** Use Fejér's theorem to deduce Weierstrass's trigonometric approximation theorem, and use that to give another proof of

1. The completeness of the orthonormal system $\{e^{2\pi inx}\}_{n \in \mathbb{Z}}$.

2. Weierstrass's polynomial approximation theorem.

**Exercise 4.5.16.** The ***Poisson kernel*** is the family of functions $\{P_r(x)\}_{r \in (0,1)}$ on $[-\pi, \pi]$ given by

$$P_r(x) = \sum_{n \in \mathbb{Z}} r^{|n|} e^{inx}.$$

1. Prove that $P_r * f(x) = \sum_{n \in \mathbb{Z}} r^{|n|} \hat{f}(n) e^{inx}$.

2. Use the formula for the sum of a geometric series in order to get the following more manageable formula for $P_r(x)$:

$$P_r(x) = \text{Re}\left(\frac{1 + re^{ix}}{1 - re^{ix}}\right) = \frac{1 - r^2}{1 - 2r\cos x + r^2}.$$

3. Prove that $P_r * f \xrightarrow{r \to 1} f$ uniformly for every $f \in C([-\pi, \pi])$ such that $f(-\pi) = f(\pi)$. (**Hint:** show that the Poisson kernel enjoys the four properties listed in Lemma 4.3.1 that the Fejér kernel enjoys; that is, the Poisson kernel is a summability kernel.)

**Exercise 4.5.17.** Fourier series can be useful for solving certain kinds of differential equations (this was, in fact, Fourier's motivation for considering them). Here is an example. *The Dirichlet problem for the heat equation on a wire* is the partial differential equation with boundary conditions

$$u_t(x,t) = cu_{xx}(x,t) \quad , \quad x \in [0,L], \ t > 0,$$
$$u(x,0) = f(x) \quad , \quad x \in [0,L],$$
$$u(0,t) = 0 = u(L,t) \quad , \quad t > 0.$$

The problem is to find a function $u \in C^2([0,L] \times [0,\infty))$ that satisfies the conditions above, where $f$ is a given function satisfying $f(0) = f(L) = 0$. This problem has a physical interpretation: $u(x,t)$ represents the temperature at the point $x$ along a one-dimensional wire (of length $L$) at the time $t$. The constant $c$ is a positive number that depends on physical considerations (e.g., the particular material the wire is made of). The function $f(x)$ is the distribution of the temperature along the wire at time 0, and we assume it is known. Assuming that the temperature of the wire at the endpoints is held at a fixed temperature 0 (this is what the conditions $u(0,t) = u(L,t)$ mean), we want to understand how the temperature distribution evolves with time, given that at time $t = 0$ it is equal to $f(x)$.

The problem can be solved as follows. Assuming that $u \in C^2$ and $u(0,t) = u(L,t) = 0$, we know that for every $t \geq 0$, $u(x,t)$ can be developed into a sine series: $u(x,t) = \sum_{n=1}^{\infty} b_n(t) \sin\left(\frac{n\pi x}{L}\right)$. Allow yourself to proceed formally, to find a representation for $u(x,t)$ in terms of $f$. Show that if $f(0) = f(L) = 0$ and if $f \in C^1([0,L])$, then the function $u$ thus found is a true solution of the problem. In particular, find $u$ when $f(x) = x(x - L)$.

Discuss: why did we choose to work with sine series? What would have gone differently if we would have chosen, say, to develop $u(x,t)$ into a cosine series?

# Chapter 5

# Bounded linear operators on Hilbert space

Recall that a map $T : V \to W$ between two vector spaces $V$ and $W$ is said to be a *linear transformation* if

$$T(\alpha u + v) = \alpha T(u) + T(v)$$

for all $\alpha$ in the field and all $u, v \in V$. The theory of linear operators on infinite dimensional spaces becomes really interesting only when additional topological assumptions are made. In this book, we will only consider the boundedness and the compactness assumptions. We begin by discussing bounded operators on Hilbert spaces.

## 5.1   Bounded operators

**Definition 5.1.1.** A linear transformation $T : H \to K$ mapping between two inner product spaces is said to be *bounded* if the *operator norm* $\|T\|$ of $T$, defined as

$$\|T\| = \sup_{\|h\|=1} \|Th\|,$$

satisfies $\|T\| < \infty$. An immediate observation is that for all $h \in H$ we have

$$\|Th\| \leq \|T\|\|h\|.$$

A bounded linear transformation is usually referred to as a *bounded operator*, or simply as an *operator*. The set of bounded operators between two inner product spaces $H$ and $K$ is denoted by $B(H, K)$, and the notation for the space $B(H, H)$ is usually abbreviated to $B(H)$.

**Example 5.1.2.** Let $H = \mathbb{C}^n$ be a finite dimensional Hilbert space (with the standard inner product). Every linear transformation on $H$ is of the form $T_A$ given by $T_A(x) = Ax$, where $A \in M_n(\mathbb{C})$ is an $n \times n$ matrix. By elementary analysis, $T_A$ is continuous, and by compactness of the unit sphere $S = \{x \in \mathbb{C}^n : \|x\| = 1\}$, $T_A$ is bounded.

Alternatively, one can find an explicit bound for $\|T_A\|$ by using the definition. Let $x \in H$. Then

$$\|Ax\|^2 = \sum_{i=1}^{n} |\sum_{j=1}^{n} a_{ij}x_j|^2.$$

By the Cauchy-Schwarz inequality, the right-hand side is less than

$$\sum_{i=1}^{n} \left( \sum_{j=1}^{n} |a_{ij}|^2 \right) \|x\|^2 = \|x\|^2 \sum_{i,j=1}^{n} |a_{ij}|^2.$$

Thus $\|T_A\| \leq \sqrt{\sum_{i,j=1}^{n} |a_{ij}|^2}$.

**Exercise 5.1.3.** Show that the above bound for $\|T_A\|$ may be attained in some cases, but that in general it is not sharp (saying that the bound is "not sharp" means that for some matrix $A$, the norm of the operator $T_A$ will be strictly smaller than the bound: $\|T_A\| < \sqrt{\sum_{i,j=1}^{n} |a_{ij}|^2}$.)

**Proposition 5.1.4.** *For a linear transformation $T : H \to K$ mapping between two inner product spaces, the following are equivalent:*

1. *$T$ is bounded.*

2. *$T$ is continuous.*

3. *$T$ is continuous at some $h_0 \in H$.*

*Proof.* If $T$ is bounded and $h_n \to h$, then

$$\|Th_n - Th\| = \|T(h_n - h)\| \leq \|T\|\|h_n - h\| \to 0,$$

so $T$ is continuous. If $T$ is continuous, then obviously it is continuous at every point in $H$.

Suppose that $T$ is unbounded; we will show that it is discontinuous at every point. If $T$ is unbounded, then we can find $h_n \to 0$ such that $\|Th_n\| \geq 1$. But then for every $h_0 \in F$, we have that $h_0 + h_n \to h_0$, while

$$\|Th_0 - T(h_0 + h_n)\| = \|Th_n\| \geq 1,$$

thus $T$ is not continuous at $h_0$. $\square$

In this book, we will almost exclusively deal with bounded operators, so we will not bother using the adjective "bounded" over and over again. The reader should keep in mind that unbounded linear operators defined on linear subspaces of Hilbert spaces are of great importance and are studied a lot (on the other hand, unbounded operators defined on the entire Hilbert space $H$ are hardly ever of any interest).

**Exercise 5.1.5.** Prove that there exists an unbounded linear operator on $\ell^2$.

A simple and useful way of constructing bounded operators on Hilbert spaces is by applying the following result.

**Proposition 5.1.6.** *Let $D$ be a dense linear subspace of a Hilbert space $H$. Let $K$ be another Hilbert space, let $T : D \to K$ be a linear transformation, and suppose that*

$$\sup\{\|Tx\| : x \in D, \|x\| = 1\} = M < \infty.$$

*Then there is a unique bounded operator $\tilde{T}$ defined on $H$ such that $\tilde{T}|_D = T$ and $\|\tilde{T}\| = M$. Moreover, if $\|Tx\| = \|x\|$ for all $x \in D$, then $\|\tilde{T}h\| = \|h\|$ for all $h \in H$.*

*Proof.* Let $h \in H$. To define $Th$, we find a sequence $D \ni x_n \to h \in H$, and we note that $\{Tx_n\}$ is a Cauchy sequence. Hence $\lim_{n\to\infty} Tx_n$ exists and we define $\tilde{T}h = \lim_{n\to\infty} Tx_n$. This limit is well-defined since if we have another sequence $D \ni y_n \to h$, then the sequence $x_1, y_1, x_2, y_2, \ldots$ also converges to $h$, thus $Tx_1, Ty_1, Tx_2, Ty_2, \ldots$ is a Cauchy sequence. In particular, if $h \in D$ then $\tilde{T}h = Th$.

$\tilde{T}$ is clearly the unique continuous map on $H$ that extends $T$, and by linearity of $T$ and continuity of the algebraic operations, $\tilde{T}$ is also linear.

The final assertion follows from continuity of the norm. $\square$

**Example 5.1.7.** Let $H = L^2[0,1]$, and let $g \in C([0,1])$. We define the **multiplication** operator $M_g : L^2[0,1] \to L^2[0,1]$ by

$$M_g f = gf.$$

The operator $M_g$ is clearly defined and bounded on the dense subspace $C([0,1])$, so one may use Proposition 5.1.6 to extend $M_g$ to a bounded operator on all of $L^2[0,1]$, and it is not difficult to compute $\|M_g\| = \|g\|_\infty$. If one knows measure theory, then this extension procedure is not needed (since for $f \in L^2[0,1]$, $f$ is defined almost everywhere, so $gf$ is defined almost everywhere). On the other hand, the extension argument allows us to extend the algebraic operation of multiplication-by-a-continuous-function from $C([0,1])$ to all of $L^2[0,1]$, strengthening our claim that $L^2[0,1]$, as we defined it, can be considered as a space of functions.

**Definition 5.1.8.** For every $T \in B(H,K)$, we define the **kernel** of $T$ to be the space

$$\ker(T) = \{h \in H : Th = 0\}.$$

The **image** of $T$ is defined to be the space

$$\operatorname{Im}(T) = \{Th : h \in H\}.$$

The kernel is sometimes referred to as the **null space** and the image is sometimes called the **range**. The kernel of a bounded operator is always closed. The image of a bounded operator might be nonclosed.

**Proposition 5.1.9.** *Let $S, T \in B(H, K)$. Then $S = T$ if and only if $\langle Sh, k \rangle = \langle Th, k \rangle$ for all $h \in H$ and $k \in K$. In case $H = K$, then $S = T$ if and only if $\langle Sh, h \rangle = \langle Th, h \rangle$ for all $h \in H$.*

**Exercise 5.1.10.** Prove the proposition, and show that the second part fails for real Hilbert spaces.

---

## 5.2    Linear functionals and the Riesz representation theorem

**Definition 5.2.1.** Let $H$ be an inner product space. A linear operator $\Phi : H \to \mathbb{C}$ is said to be a **linear functional**. The space of all bounded linear functionals on $H$ is denoted by $H^*$.

**Example 5.2.2.** Let $H$ be an inner product space, and fix $g \in H$. We define a linear functional $\Phi_g$ on $H$ as follows:

$$\Phi_g(h) = \langle h, g \rangle , \quad h \in H.$$

By the linearity of the inner product in the first component, $\Phi_g$ is a linear functional. For every $h \in H$, Cauchy-Schwarz implies that $|\Phi_g(h)| = |\langle h, g \rangle| \leq \|h\| \|g\|$, thus $\|\Phi_g\| \leq \|g\|$. When $g \neq 0$, we apply $\Phi_g$ to the unit vector $u = g/\|g\|$ and obtain $\Phi_g(u) = \|g\|$; thus $\|\Phi_g\| = \|g\|$.

In fact, in a Hilbert space, the above example covers all the possibilities.

**Theorem 5.2.3** (Riesz representation theorem). *If $\Phi$ is a bounded linear functional on a Hilbert space $H$, then there exists a unique $g \in H$, such that $\Phi(h) = \langle h, g \rangle$ for all $h \in H$. In addition, $\|g\| = \|\Phi\|$.*

*Proof.* If $\Phi = 0$, then we choose $g = 0$ and the proof is complete, so assume that $\Phi \neq 0$. Then $\ker \Phi$ is a closed subspace which is not $H$. By Theorem 3.2.10 and the following remark, we have $H = \ker \Phi \oplus \ker \Phi^\perp$. Since the restriction of $\Phi$ to $\ker \Phi^\perp$ is a linear map with trivial kernel that maps onto $\mathbb{C}$, the dimension of $\ker \Phi^\perp$ is equal to 1. Choose a unit vector $u \in \ker \Phi^\perp$. We claim that $g = \overline{\Phi(u)}u$ does the trick. Indeed, let $h \in H$. Let $h = m + n$ be the decomposition of $h$ with respect to $H = \ker \Phi^\perp \oplus \ker \Phi$. By the formula of orthogonal projection onto a subspace (Proposition 3.2.18), $m = \langle h, u \rangle u$. Then we have

$$\Phi(h) = \Phi(m) + \Phi(n) = \Phi(\langle h, u \rangle u) = \langle h, u \rangle \Phi(u) = \langle h, g \rangle.$$

The equality $\|g\| = \|\Phi\|$ has been worked out in the above example. Finally, $g$ is unique, because if $g'$ is another candidate for this role, then $\langle h, g - g' \rangle = \Phi(h) - \Phi(h) = 0$ for all $h \in H$, whence $g = g'$ (because, in particular, $\langle g - g', g - g' \rangle = 0$). $\qquad\square$

**Exercise 5.2.4.** Where did we use the fact that $H$ is a Hilbert space? Where did we use the fact that $\Phi$ is bounded? Construct examples showing that neither of these assumptions can be dropped.

**Exercise 5.2.5.** Prove that a linear functional on a Hilbert space is bounded if and only if its kernel is closed.

---

## 5.3   *The Dirichlet problem for the Laplace operator

The material in this section is supplementary, and the reader may skip to the next section without worries. Here we will illustrate an application of the Riesz representation theorem to the problem of finding solutions to partial differential equations (PDEs). We will outline an example involving the Laplace operator. For simplicity, we consider real-valued functions in two variables — this is already a case of much importance, in physics and in pure mathematics. For a detailed introduction to the use of functional analytic methods in PDEs, the reader may consult [4].

Recall that the **Laplace operator** $\Delta$ is the partial differential operator defined on smooth functions $u = u(x, y)$ by

$$\Delta u = u_{xx} + u_{yy}.$$

We are using the familiar shorthand notation for partial differentiation, where $u_x = \frac{\partial u}{\partial x}$, $u_{xx} = \frac{\partial^2 u}{\partial x^2}$, etc.

Suppose that we are given a bounded domain (connected open set) $D \subset \mathbb{R}^2$ together with a continuous real-valued function $f \in C(\overline{D})$, and consider the following boundary value problem:

$$\Delta u = f \quad \text{in } D \tag{5.1}$$

$$u\big|_{\partial D} = 0 \quad \text{on } \partial D \tag{5.2}$$

Here and below, $\partial D$ denotes the boundary of $D$ (that is $\partial D$ is the set of all points $x \in \mathbb{R}^2$, such that every neighborhood of $x$ intersects both $D$ and its complement). This problem is called the *Dirichlet problem for the Laplace operator*. We seek a solution to this problem. A **strong solution** (also called **classical solution**) for this problem is a function $u \in C(\overline{D}) \cap C^2(D)$ such that $\Delta u = f$ in $D$ and $u \equiv 0$ on $\partial D$. It will be convenient to denote

$$C_0^2(\overline{D}) = \{u \in C(\overline{D}) \cap C^2(D) : u\big|_{\partial D} \equiv 0\}.$$

When $D$ is a disc or a rectangle, there are methods to write down a solution $u$ in terms of $f$. However, it is desirable to have a general theory of solvability of the Dirichlet problem, that works for as wide as possible a class of bounded domains $D$. One approach is to recast the problem within the framework of Hilbert spaces.

Let $C_0^1(\overline{D})$ be the space of continuously differentiable real-valued functions in $\overline{D}$, which vanish on the boundary of $D$. We define an inner product on $C_0^1(\overline{D})$

$$\langle u, v \rangle_{H_0^1} = \int_D \nabla u \cdot \nabla v = \int_D (u_x v_x + u_y v_y) dx dy.$$

(We will henceforth omit the variables from integration, and write $\int_D f$ for $\int_D f(x,y) dx dy$). This is indeed an inner product: it is definite because if $\|u\|_{H_0^1} = 0$, then $u$ must be constant, and the boundary condition then implies $u = 0$. We write $\langle \cdot, \cdot \rangle_{H_0^1}$ for the inner product and $\| \cdot \|_{H_0^1}$ for the norm, since we wish to avoid confusion with the $L^2$ inner product

$$\langle u, v \rangle_{L^2} = \int_D uv.$$

Let $H_0^1(D)$ denote the completion of the space $C_0^1(\overline{D})$ with respect to the inner product $\langle \cdot, \cdot \rangle_{H_0^1}$.

**Remark 5.3.1.** Like $L^2(D)$, the space $H_0^1(D)$ can be interpreted as a space of functions, and is a member of a large and important family of spaces called *Sobolev spaces*. One can define a space $H^1(D)$ that consists of all functions on $D$, for which the first-order derivatives $u_x$ and $u_y$ exist in some weak sense, and such that $u_x, u_y \in L^2(D)$. The subscript 1 therefore stands for first-order derivatives, and a space $H^k(D)$ of functions with $k$th order derivatives in $L^2(D)$ can also be defined. $H_0^1(D)$ is then the subspace of $H^1(D)$ consisting of functions that "vanish on the boundary". Sobolev spaces are indispensable for significant parts of the study of PDEs; the reader is referred to almost any comprehensive textbook on the modern theory of PDEs (such as [10]) for more on these spaces and how they are used.

We define also the space of **test functions** to be the space $C_c^\infty(D)$ of all infinitely differentiable real-valued functions that vanish on a neighborhood of the boundary. It is useful to introduce the following concept.

**Definition 5.3.2.** A **weak solution** to the boundary value problem (5.1)-(5.2) is an element $u \in H_0^1(D)$ such that

$$\langle u, \phi \rangle_{H_0^1} = - \int_D f\phi \qquad (5.3)$$

for every $\phi \in C_c^\infty(D)$.

A couple of remarks are in order. First, if $u$ is in $C_0^2(\overline{D})$ and is a strong solution for (5.1)-(5.2), then it is also a weak solution. Indeed, using integration by parts, it is easy to see that a strong solution $u$ satisfies

$$\langle u, \phi \rangle_{H_0^1} = \int_D u_x \phi_x + u_y \phi_y = - \int_D (\Delta u)\phi = - \int_D f\phi.$$

On the other hand, if $u \in H_0^1(D)$ is a weak solution, and $u$ also happens to lie in the subspace $C_0^2(\overline{D})$, then $u$ is in fact a strong solution. To see this, we use integration by parts, and, assuming that $u$ is a weak solution, we get

$$\int_D (\Delta u)\phi = - \int_D u_x \phi_x + u_y \phi_y = -\langle u, \phi \rangle_{H_0^1} = \int_D f\phi,$$

so $\int_D (\Delta u)\phi = \int_D f\phi$ for all $\phi \in C_c^\infty(D)$. Invoking the following lemma, we conclude that $\Delta u = f$.

**Lemma 5.3.3.** *If $f, g \in C(D)$ and $\int_D f\phi = \int_D g\phi$ for all $\phi \in C_c^\infty(D)$, then $f \equiv g$.*

*Proof.* If $f(x_0) \neq g(x_0)$ at some $x_0 \in D$, then there is some neighborhood $B_r(x_0)$ of $x_0$ where $f - g$ is nonzero and of constant sign. Now choose the test function $\phi$ to be positive at $x_0$, nonnegative everywhere, and 0 outside $B_r(x_0)$. Integrating against this test function $\phi$, we get $\int_D (f - g)\phi \neq 0$ (for the existence of such a test function, see the proof of Theorem 12.5.1). □

In the above discussion we have proved: *a function $u \in C_0^2(\overline{D})$ is a strong solution to (5.1)-(5.2) if and only if it is a weak solution.* Thus, the notion of a weak solution that we have introduced is a reasonable one. The proof of the following theorem makes use of the Riesz representation theorem to establish the existence of a weak solution.

**Theorem 5.3.4.** *For every $f \in C(\overline{D})$, the problem (5.1)-(5.2) has a weak solution.*

*Proof.* We define a linear functional

$$\Phi : C_0^1(\overline{D}) \to \mathbb{R}$$

by

$$\Phi(v) = - \int_D fv. \tag{5.4}$$

The idea is to use the Riesz representation theorem, which tells us that in order to obtain the existence of a weak solution, it suffices to show that $\Phi$ extends to a bounded linear functional on $H_0^1(D)$. By Proposition 5.1.6, this boils town to showing that this $\Phi$ is bounded on $C_0^1(\overline{D})$ with respect to the norm $\| \cdot \|_{H_0^1}$. We pause to state and prove this fact as a lemma.

**Lemma 5.3.5.** *There exists a constant $C > 0$, such that for every $v \in C_0^1(\overline{D})$,*

$$\left| \int_D fv \right| \le C \|v\|_{H_0^1}.$$

*Hence, Equation (5.4) gives rise to a bounded linear functional $\Phi$ on $H_0^1(D)$.*

*Proof.* By the Cauchy-Schwarz inequality in $L^2(D)$,

$$\left| \int_D fv \right| \le \|f\|_{L^2} \|v\|_{L^2}.$$

Therefore, it suffices to prove that

$$\|v\|_{L^2} \le C \|v\|_{H_0^1} \quad \text{for every } v \in C_0^1(\overline{D}), \tag{5.5}$$

for some constant $C$. Since $D$ is bounded, we can assume that it is contained in some square $[-R, R] \times [-R, R]$. Considering $v \in C_0^1(\overline{D})$ as a function defined on all $[-R, R] \times [-R, R]$ and vanishing outside $D$, we have for every $x, y \in D$,

$$|v(x, y)| \le \int_{-R}^{x} |v_x(t, y)| dt \le (2R)^{1/2} \left( \int_{-R}^{R} |v_x(t, y)|^2 dt \right)^{1/2}.$$

Squaring the above inequality and integrating over $y$ we obtain, for every fixed $x \in [-R, R]$,

$$\int_{-R}^{R} |v(x, y)|^2 dy \le 2R \int_{-R}^{R} \int_{-R}^{R} |v_x(t, y)|^2 dt dy \le 2R \|v\|_{H_0^1}^2.$$

Integrating over $x$ gives

$$\int_{-R}^{R} \int_{-R}^{R} |v(x, y)|^2 dx dy \le \int_{-R}^{R} 2R \|v\|_{H_0^1}^2 = (2R)^2 \|v\|_{H_0^1}^2,$$

which gives the required inequality (5.5). $\square$

Now we can complete the proof of the theorem. By the Riesz representation theorem, there exists a unique $u \in H_0^1(D)$ such that

$$\langle u, v \rangle_{H_0^1} = \Phi(v)$$

for all $v \in H_0^1(D)$. In pacticular, this $u$ is an element in $H_0^1(D)$ that satisfies

$$\langle u, \phi \rangle_{H_0^1} = -\int_D f\phi \quad \text{for all } \phi \in C_c^\infty(D).$$

In other words, $u$ is a weak solution. $\square$

**Exercise 5.3.6.** In the above theorem we obtained the existence of a weak solution $u$, but we did not say anything about uniqueness. In fact, the weak solution is also unique. Do you see what argument needs to be added to the proof in order to establish uniqueness? (The student who sees what is missing, but does not know how to fill in the missing details, might want to return to this exercise after reading the proof of Theorem 12.5.1.)

We started out by stating the Dirichlet problem for the Laplace operator, and our discussion has led us to conclude that there exists a weak solution to that problem. This is interesting, and it might be useful, but it does not answer the original problem of finding a strong solution.

Remarkably, it turns out that obtaining a weak solution is a significant step towards solving the classical problem. One can show that if $f$ and $D$ are nice enough (for example, if $f \in C^1(\overline{D})$, and if $D$ has a smooth boundary), then the weak solution $u$ must be an element of the space $C_0^2(\overline{D})$. Hence, by the discussion preceding the theorem, a weak solution is a strong solution. This is a nontrivial "hard analytical" step, and is beyond the scope of this book. It is important to point out that the proof that a weak solution must be in $C_0^2(\overline{D})$ (an hence is a strong solution) is best carried out within the framework of Lebesgue measure theory, which we are systematically avoiding in this book.

**Exercise 5.3.7.** Prove directly (without relying on uniqueness of the weak solution) that a strong solution is unique.

We now summarize. In this section, we presented a typical application of the Riesz theorem to showing existence of solutions to PDEs. In the setting of the Dirichlet problem for the Laplace operator in the plane, this consists of the following steps:

1. Define a cleverly chosen linear functional $\Phi$ on $H_0^1(D)$, and prove that it is bounded.

2. Use the Riesz theorem to find $u$ such that $\Phi(v) = \langle u, v \rangle$ for all $v \in H_0^1(D)$, and recognize that (thanks to the clever choice of $\Phi$) $u$ is a weak solution.

3. Use hard analysis to obtain that $u$ lies in a nice subspace, say $u \in C_0^2(\overline{D})$, thus $u$ is a strong solution, and the problem is solved.

It is important to note that there are two nontrivial analytical things to prove; the boundedness of $\Phi$ and the fact that $u$ lies in $C_0^2(\overline{D})$. Without these analytical steps, the functional analytic step (2) gives us very little.

**Remark 5.3.8.** Weak solutions are usually defined in a slightly different way from the way we defined them. If we think of $L^2(D)$ and $H_0^1(D)$ as spaces consisting of measurable functions, then the inequality (5.5) shows that $H_0^1(D) \subset L^2(D)$. Therefore, if $u \in H_0^1(D)$ is a weak solution, then in particular $u, u_x, u_y \in L^2(D)$. Integration by parts is therefore applicable, and (5.3) can be rewritten as

$$- \int f\phi = \langle u, \phi \rangle_{H_0^1} = \int u_x \phi_x + u_y \phi_y = - \int u \Delta \phi.$$

In other words, $u$ is a weak solution if and only if

$$\langle u, \Delta \phi \rangle_{L^2} = \int_D f\phi \quad \text{for all } \phi \in C_c^\infty(D).$$

## 5.4   The adjoint of a bounded operator

**Proposition 5.4.1** (Existence of the adjoint). *Let $H$ and $K$ be Hilbert spaces. For every $T \in B(H, K)$ there exists a unique $S \in B(K, H)$ such that*

$$\langle Th, k \rangle = \langle h, Sk \rangle$$

*for all $h \in H$ and $k \in K$.*

*Proof.* Fix $k \in K$, and consider the functional $\Phi(h) = \langle Th, k \rangle$. It is clear that $\Phi$ is linear, and the estimate

$$|\Phi(h)| \leq \|Th\|\|k\| \leq \|T\|\|k\|\|h\|$$

shows that $\Phi$ is bounded and $\|\Phi\| \leq \|T\|\|k\|$. By the Riesz representation theorem (Theorem 5.2.3), there exists a unique element in $H$, call it $Sk$, such that $\langle h, Sk \rangle = \Phi(h) = \langle Th, k \rangle$ for all $h$. What remains to be proved is that the mapping $k \mapsto Sk$ is a bounded linear map.

Linearity is easy. For example, if $k_1, k_2, \in K$, then for every $h \in H$

$$\langle h, S(k_1 + k_2) \rangle = \langle Th, k_1 + k_2 \rangle = \langle Th, k_1 \rangle + \langle Th, k_2 \rangle = \langle h, Sk_1 + Sk_2 \rangle,$$

so $S(k_1 + k_2) = Sk_1 + Sk_2$. As for boundedness, Theorem 5.2.3 also tells us that $\|Sk\| = \|\Phi\| \leq \|T\|\|k\|$, thus $S$ is bounded and $\|S\| \leq \|T\|$. The uniqueness of $S$ also follows, from the unique choice of $Sk$ given $k$.   $\square$

**Definition 5.4.2.** For every $T \in B(H, K)$, we let $T^*$ denote the unique operator for which $\langle Th, k \rangle = \langle h, T^*k \rangle$ for all $h \in H$ and $k \in K$. The operator $T^*$ is called **the adjoint** of $T$.

**Example 5.4.3.** Let $H = \mathbb{C}^n$, let $A \in M_n(\mathbb{C})$, and let $T_A \in B(H)$ be multiplication by $A$. If $A = (a_{ij})$, then let $A^*$ be the matrix with $ij$th entry $\overline{a_{ji}}$. Then $T_A^* = T_{A^*}$. This should be well known from a course in linear algebra, but is also immediate from the definitions:

$$\langle T_A x, y \rangle = y^* A x = (A^* y)^* x = \langle x, T_{A^*} y \rangle,$$

for all $x, y \in \mathbb{C}^n$.

**Example 5.4.4.** Let $H = L^2[0,1]$, and let $g \in C([0,1])$. Then $M_g^* = M_{\bar{g}}$. To see this, we compute for all $h, k \in C([0,1])$

$$\langle M_g h, k \rangle = \int_0^1 g(t)h(t)\overline{k(t)}dt = \langle h, M_{\bar{g}}k \rangle,$$

so $\langle h, M_g^* k \rangle = \langle h, M_{\bar{g}}k \rangle$. Because $M_g^*$ and $M_{\bar{g}}$ are bounded, and because the inner product is continuous, it follows that $\langle h, M_g^* k \rangle = \langle h, M_{\bar{g}}k \rangle$ holds for all $h, k \in L^2[0,1]$, thus $M_g^* = M_{\bar{g}}$.

**Proposition 5.4.5.** *Let $S, T$ be bounded operators between Hilbert spaces. The following hold:*

1. $(T^*)^* = T$.

2. $\|T\| = \|T^*\|$.

3. $(aS + bT)^* = \bar{a}S^* + \bar{b}T^*$ *for* $a, b \in \mathbb{C}$.

4. $(ST)^* = T^*S^*$.

5. $\|T\|^2 = \|T^*T\|$.

6. *If $T$ is bijective and $T^{-1}$ is bounded, then $T^*$ is invertible and $(T^*)^{-1} = (T^{-1})^*$.*

**Exercise 5.4.6.** Prove Proposition 5.4.5.

**Remark 5.4.7.** We will later learn that when $T \in B(H, K)$ is invertible, then $T^{-1}$ is automatically bounded (see Theorem 8.3.5).

**Proposition 5.4.8.** *Let $T \in B(H, K)$. We have the following:*

1. $\operatorname{Im} T^\perp = \ker T^*$.

2. $(\operatorname{Im} T^*)^\perp = \ker T$.

3. $\ker T^\perp = \overline{\operatorname{Im} T^*}$.

4. $(\ker T^*)^\perp = \overline{\operatorname{Im} T}$.

*Proof.* Items (1) and (2) are equivalent, and so are (3) and (4) (thanks to the first item in Proposition 5.4.5).

To prove (1), let $k \in K$. Then $T^*k = 0$ if and only if for all $h$, $0 = \langle T^*k, h \rangle = \langle k, Th \rangle$, and this happens if and only if $k \perp \operatorname{Im} T$. To prove (4), apply $\perp$ to both sides of (1) to obtain $(\ker T^*)^\perp = \operatorname{Im} T^{\perp\perp} = \overline{\operatorname{Im} T}$. $\square$

## 5.5   Special classes of operators

**Definition 5.5.1.** An operator $T \in B(H)$ is said to be

- *normal* if $T^*T = TT^*$,

- *self-adjoint* if $T = T^*$,

- *positive* if $\langle Th, h \rangle \geq 0$ for all $h \in H$.

An operator $T \in B(H, K)$ is said to be

1. *contractive*, or a *contraction*, if $\|Th\| \leq \|h\|$ for all $h \in H$,

2. *isometric*, or an *isometry*, if $\|Th\| = \|h\|$ for all $h \in H$,

3. *unitary* if it is a bijection satisfying $\langle Tg, Th \rangle = \langle g, h \rangle$ for all $g, h \in H$.

**Example 5.5.2.** Let $I$ denote that identity operator on $H$, that is $Ih = h$ for all $h \in H$. Then $I$ belongs to all of the above categories.

**Proposition 5.5.3.** $T \in B(H)$ *is self-adjoint if and only if* $\langle Th, h \rangle \in \mathbb{R}$ *for all* $h \in H$.

**Proposition 5.5.4.** *For* $T \in B(H)$, *the following are equivalent:*

1. *$T$ is an isometry.*

2. *$\langle Tg, Th \rangle = \langle g, h \rangle$ for all $g, h \in H$.*

3. *$T^*T = I$.*

**Proposition 5.5.5.** *Let* $T \in B(H)$. *$T$ is unitary if and only if*

$$T^*T = TT^* = I.$$

**Exercise 5.5.6.** Prove the above propositions. Which of them fail when the field of scalars $\mathbb{C}$ is replaced by $\mathbb{R}$?

**Example 5.5.7.** Let $M$ be a closed subspace of a Hilbert space. Then the orthogonal projection $P_M$ onto $M$ is positive and self-adjoint. Indeed, if $h \in H$, write $h = m + n$ where $m \in M, n \in M^\perp$. Then $\langle P_M h, h \rangle = \langle m, m + n \rangle = \langle m, m \rangle \geq 0$, so it is positive. By Proposition 5.5.3, $P_M$ is self-adjoint.

Alternatively, if $h = m + n$ and $h' = m' + n'$ are decompositions of two vectors according to $H = M \oplus M^\perp$, then $\langle P_M h, h' \rangle = \langle m, m' \rangle = \langle h, P_M h' \rangle$, thus $P_M = P_M^*$, so $P_M$ is self-adjoint by definition (why was it important to give a different proof?).

**Example 5.5.8.** Let $S$ be the operator on $\ell^2$ given by

$$S(x_1, x_2, x_3, \dots,) = (0, x_1, x_2, \dots).$$

The operator $S$ is called the **unilateral shift** or the **forward shift** (or simply the **shift**). It is easy to see that $S$ is an isometry. A computation shows that $S^*$ is given by $S^*(x_1, x_2, x_3, \dots) = (x_2, x_3, \dots)$. The operator $S^*$ is called **the backward shift**. Further computations show that $S^*S = I$, and $SS^* = P_M$, where $M = \{(1, 0, 0, 0, \dots)\}^{\perp}$. So $S$ is not normal.

---

## 5.6   Matrix representation of operators

Consider the Hilbert space $\ell^2 = \ell^2(\mathbb{N})$. Suppose that we are given an infinite matrix $A = (a_{ij})_{i,j \in \mathbb{N}}$ such that

$$\sum_{i,j=0}^{\infty} |a_{ij}|^2 < \infty. \tag{5.6}$$

Since every row of $A$ is in $\ell^2$, we can multiply $A$ with elements of $\ell^2$ (considered as infinite column vectors) as follows: the vector $Ax$ is the sequence where the $i$th element is

$$(Ax)_i = \sum_{j=0}^{\infty} a_{ij} x_j$$

for all $i \in \mathbb{N}$. As in Example 5.1.2, we find that $Ax \in \ell^2$, and in fact that the operator $T_A : x \mapsto Ax$ is a bounded operator with $\|T_A\| \leq \sqrt{\sum_{i,j=0}^{\infty} |a_{ij}|^2}$. Indeed,

$$\|Ax\|^2 = \sum_{i=0}^{\infty} \left| \sum_{j=0}^{\infty} a_{ij} x_j \right|^2 \leq \|x\|^2 \sum_{i,j=0}^{\infty} |a_{ij}|^2. \tag{5.7}$$

It is common practice to slightly abuse notation and use the symbol $A$ to denote both the matrix $A$ as well as the operator $T_A$.

There are infinite matrices which do not satisfy (5.6) that give rise to bounded operators; for example, the "infinite identity matrix" gives rise by multiplication to the identity operator on $\ell^2(\mathbb{N})$. There is no known condition on the coefficients of an infinite matrix $A$ (such as (5.6)) that guarantees that the matrix multiplication by $A$ gives rise to a well-defined bounded operator on $\ell^2$.

On the other hand, every bounded operator on a Hilbert space has a "matrix representation".

**Definition 5.6.1.** Let $H$ be a Hilbert space with an orthonormal basis $\{f_i\}_{i \in I}$, and let $T \in B(H)$. The *representing matrix of $T$ with respect to the basis* $\{f_i\}_{i \in I}$ is the array $[T] = (t_{ij})_{i,j \in I}$ given by

$$t_{ij} = \langle Tf_j, f_i \rangle, \quad i, j \in I.$$

We also sometimes write $[T]_{ij}$ for the $ij$th element $t_{ij}$.

For every $h = \sum_{j \in I} \langle h, f_j \rangle f_j \in H$, the operation of $T$ on $h$ is given by

$$Th = \sum_{i \in I} \left( \sum_{j \in I} t_{ij} \langle h, f_j \rangle \right) f_i.$$

To see this, we compute the $i$th generalized Fourier coefficient of $Th$:

$$\langle Th, f_i \rangle = \left\langle T \sum_{j \in I} \langle h, f_j \rangle f_j, f_i \right\rangle = \sum_{j \in I} \langle h, f_j \rangle \langle Tf_j, f_i \rangle,$$

as claimed.

A perhaps more concrete way to see how $T$ works is as follows. Let $\ell^2(I)$ be the Hilbert space of "sequences" indexed by $I$ (recall Exercise 3.4.7). Let $\{e_i\}_{i \in I}$ denote the standard basis of $\ell^2(I)$, that is, $e_i = (\delta_{ij})_{j \in I}$ (i.e., $e_i$ has 1 in the $i$th slot and zeros elsewhere). The map $U : f_i \mapsto e_i$ extends to a unitary map from $H$ onto $\ell^2(I)$, given by

$$Uh = U \left( \sum_{j \in I} \langle h, f_j \rangle f_j \right) = \sum_{j \in I} \langle h, f_j \rangle e_j = (\langle h, f_j \rangle)_{j \in I}.$$

Multiplying the "infinite column vector" $(\langle h, f_j \rangle)_{j \in I}$ by the matrix $[T]$ gives the column vector which has at its $i$th place the number $\sum_{j \in I} t_{ij} \langle h, f_j \rangle$, that is

$$[T]Uh = \left( \sum_{j \in I} t_{ij} \langle h, f_j \rangle \right)_{i \in I}.$$

We see that $U^*[T]Uh = Th$ for every $h \in H$, or simply

$$U^*[T]U = T.$$

The relationship between $T$ and $[T]$ is important enough that we pause to make a definition.

**Definition 5.6.2.** If $A \in B(H)$ and $B \in B(K)$, then $A$ and $B$ are said to be *unitarily equivalent* if there exists a unitary operator $U : H \to K$ such that

$$U^*BU = A.$$

To summarize, we have that every bounded operator $T$ is unitarily equivalent to the operator on $\ell^2(I)$ given by multiplication by the representing matrix of $T$. The same considerations hold for operators acting between different Hilbert spaces $H$ and $K$, or if different bases are used to represent $H$ before or after the action of $T$.

**Example 5.6.3.** Let $f \in C([0,1])$. Then one can define a convolution operator $C_f : PC[0,1] \to PC[0,1]$ by

$$C_f g(x) = f * g(x) = \int_0^1 f(x-t)g(t)dt.$$

For this definition to make sense, we extend $f$ periodically to the whole real line (with some ambiguity on the integers). By Cauchy-Schwarz, $|C_f g(x)| \leq \|f\|_2\|g\|_2$ for all $x \in [0,1]$, so $C_f$ is bounded with $\|C_f\| \leq \|f\|_2$. We can extend $C_f$ to an operator defined on $L^2[0,1]$. By Exercise 4.5.12, the representing matrix of $C_f$ with respect to the basis $\{e^{2\pi inx}\}_{n \in \mathbb{Z}}$ is given by the doubly infinite matrix that has $\hat{f}(n)$ on the $n$th place of the diagonal, and has 0s in all off diagonal entries:

$$[C_f] = \begin{pmatrix} \ddots & & & & \\ & \hat{f}(-1) & & & \\ & & \hat{f}(0) & & \\ & & & \hat{f}(1) & \\ & & & & \hat{f}(2) \\ & & & & & \ddots \end{pmatrix}.$$

**Exercise 5.6.4.** Prove that algebraic operations between operators correspond to natural matrix operations. In particular, prove that

1. $[S+T]_{ij} = [S]_{ij} + [T]_{ij}$,

2. $[ST]_{ij} = \sum_k [S]_{ik}[T]_{kj}$,

3. $[T^*]_{ij} = \overline{[T]_{ji}}$.

---

## 5.7  Additional exercises

**Exercise 5.7.1.** Prove the "Hahn-Banach extension theorem for Hilbert spaces": *Let $H$ be a Hilbert space, $M$ a subspace of $H$, and let $f$ be a bounded linear functional on $M$. Then there exists a bounded linear functional $F$ on $H$, such that $\|F\| = \|f\|$ and $F\big|_M = f$.*

**Exercise 5.7.2.** Let $g \in C([0,1])$, and let $M_g$ be the operator discussed in Examples 5.1.7 and 5.4.4. Prove that $\|M_g\| = \|g\|_\infty$ and that $M_g^* = M_{\bar{g}}$.

**Exercise 5.7.3.** Let $U$ be the operator on $\ell^2(\mathbb{Z})$ that shifts every bi-infinite sequence to the right; that is, $U(x_n)_{n\in\mathbb{Z}} = (x_{n-1})_{n\in\mathbb{Z}}$. The operator $U$ is called the **bilateral shift**.

1. Prove that $U$ is a unitary operator.

2. Find the matrix representation of $U$ with respect to the standard basis of $\ell^2(\mathbb{Z})$.

3. Prove that $U$ is unitarily equivalent to the multiplication operator $M_g$ on $L^2[0,1]$, where $g(t) = e^{2\pi it}$.

**Exercise 5.7.4.** Let $G, H$ be Hilbert spaces, and suppose that we are given subsets $\mathcal{A} \subseteq G$ and $\mathcal{B} \subseteq H$. Suppose that there is a map $\varphi : \mathcal{A} \to \mathcal{B}$.

1. Prove that there exists an isometry $T : \overline{\operatorname{span}\mathcal{A}} \to \overline{\operatorname{span}\mathcal{B}}$ such that $T(a) = \varphi(a)$ for all $a \in \mathcal{A}$, if and only if

$$\langle \varphi(a_1), \varphi(a_2) \rangle_H = \langle a_1, a_2 \rangle_G \quad \text{for all } a_1, a_2 \in \mathcal{A}.$$

2. Prove that there exists a contraction $T : \overline{\operatorname{span}\mathcal{A}} \to \overline{\operatorname{span}\mathcal{B}}$ such that $T(a) = \varphi(a)$ for all $a \in \mathcal{A}$, if and only if

$$\sum_{i,j=1}^n c_i\overline{c_j}\langle \varphi(a_i), \varphi(a_j) \rangle_H \le \sum_{i,j=1}^n c_i\overline{c_j}\langle a_i, a_j \rangle_G,$$

for all $a_1, \ldots, a_n \in \mathcal{A}$ and all $c_1, \ldots, c_n \in \mathbb{C}$.

**Exercise 5.7.5.** Prove that if $T \in B(H)$ is a positive operator, then all its eigenvalues are real and nonnegative. (Recall that a complex number $\lambda$ is an **eigenvalue** of $T$ if there exists a nonzero $h \in H$ such that $Th = \lambda h$.)

**Exercise 5.7.6.** Prove that for every $T \in B(H, K)$, the operator $T^*T$ is positive. The converse is also true, but will not be proved in full generality in this book. Can you prove the converse direction in finite dimensional spaces?

**Exercise 5.7.7.** Let $A$ be an $m \times n$ matrix, and consider $A$ also as an operator between the Hilbert space $\mathbb{C}^n$ and $\mathbb{C}^m$, with the standard inner product. Prove that the operator norm $\|A\|$ can be computed as follows:

$$\|A\| = \sqrt{\lambda_{max}(A^*A)},$$

where $\lambda_{max}(A^*A)$ denotes the largest eigenvalue of $A^*A$. Does this continue to hold true over the reals?

**Exercise 5.7.8.** Let $A = \left(\begin{smallmatrix} 1 & 2 & 0 & -2 & -1 \\ 2 & -1 & 1 & -1 & 1 \end{smallmatrix}\right)$. Compute $\|A\|$.

**Exercise 5.7.9.** Let $T \in B(H)$ such that $T^2 = T$. Such an operator is sometimes called a projection (though sometimes, in the setting of Hilbert spaces, the word *projection* is used to mean *orthogonal projection*). Prove the following statements.

1. $\operatorname{Im} T$ is closed, and every $h \in H$ has a unique decomposition $h = m + n$, where $m \in \operatorname{Im} T$ and $n \in \ker T$.

2. $T$ is an orthogonal projection onto $\operatorname{Im} T$ if and only if $T = T^*$.

3. $T$ is an orthogonal projection onto $\operatorname{Im} T$ if and only if $\|T\| \leq 1$.

**Exercise 5.7.10.** Give an example of a projection which is not an orthogonal projection. A projection $T = T^2 \in B(H)$, with $M = \operatorname{Im} T$ and $N = \ker T$, is sometimes referred to as the *projection onto $M$ parallel to $N$*. Explain why.

**Exercise 5.7.11** (Lax-Milgram theorem). Let $H$ be a real Hilbert space. A **bounded bilinear form** on $H$ is a function $B : H \times H \to \mathbb{R}$ which is linear in each of its variables separately (i.e., $B(f + g, h) = B(f, h) + B(g, h)$, $B(\alpha f, g) = \alpha B(f, g)$, and likewise in the second variable), and for which there exists some $C$ such that

$$|B(f, g)| \leq C \|f\| \|g\| \quad \text{for all } f, g \in H.$$

A bilinear form is said to be **coercive** if there exists a constant $c > 0$ such that

$$|B(f, f)| \geq c \|f\|^2 \quad \text{for all } f \in H.$$

1. Prove that for every bounded bilinear form $B$ on $H$, there exists a unique bounded operator $T_B \in B(H)$, such that

$$B(f, g) = \langle T_B f, g \rangle \quad \text{for all } f, g \in H.$$

2. Prove that if $B$ is a coercive bounded bilinear form on $H$, then for every bounded linear functional $\Phi \in H^*$, there exists a unique $u_\Phi \in H$, such that

$$B(u_\Phi, g) = \Phi(g) \quad \text{for all } g \in H.$$

Prove also that $\|u_\Phi\| \leq \frac{1}{c} \|\Phi\|$.

The second statement that you proved is called the "Lax-Milgram theorem", and it is very useful in finding weak solutions to PDEs, in a way similar to what we have done in Section 5.3.

**Exercise 5.7.12.** Let $f \in C([0, 1])$, and recall that $M_g$ denotes the multiplication operator on $L^2[0, 1]$.

1. Prove that for all $f \in L^2[0, 1]$, the Fourier coefficients of $gf$ are given by

$$\widehat{gf}(n) = \sum_{k \in \mathbb{Z}} \hat{f}(n - k)\hat{g}(k).$$

(Recall Exercise 4.5.12).

2. Find the matrix representation of the operator $M_g$ with respect to the basis $\{e^{2\pi inx}\}_{n\in\mathbb{Z}}$.

3. Let $U$ be the bilateral shift (see Exercise 5.7.3). Prove that $[M_g]U = U[M_g]$ for every $g \in C([0,1])$.

4. Let $T$ be an operator on $L^2[0,1]$ and let $[T]$ denote its matrix representation with respect to the basis $\{e^{2\pi inx}\}_{n\in\mathbb{Z}}$. Show that if $[T]U = U[T]$, then the diagonals of $T$ are constant, that is $[T]_{i,j} = [T]_{i+1,j+1}$ for all $i,j \in \mathbb{Z}$.

5. If $[T]U = U[T]$, does it necessarily follow that $[T][M_g] = [M_g][T]$ for every $g \in C([0,1])$?

**Exercise 5.7.13.** Let $M$ be a closed subspace of a Hilbert space $H$.

1. The orthogonal projection $P_M$ can be considered as an operator in $B(H)$, in which case $P_M$ is self-adjoint. But $P_M$ can also as an operator in $B(H, M)$, and then $P_M$ cannot be self-adjoint if $M \neq H$. What is the adjoint of $P_M$ when viewed this way?

2. If $T \in B(H)$, we let $T\big|_M$ denote the restriction of $T$ to $M$, given by

$$T\big|_M m = Tm \quad \text{for all } m \in M.$$

Prove that $T\big|_M \in B(M, H)$, and compute $(T\big|_M)^*$.

**Exercise 5.7.14** (Operator block matrices). Let $\{H_i\}_{i\in I}$ be a family of Hilbert spaces, and let $H = \oplus_{i\in I} H_i$ (as you defined in Exercise 3.6.8). For every $i$, let $P_i$ be the projection of $H$ onto the natural copy of $H_i$ in $H$ (living in the $i$th slot). For every $i,j$, let $T_{ij} \in B(H_j, H_i)$ be given by $T_{ij} = P_i T\big|_{H_j}$. The *operator block matrix decomposition* of $T$ with respect to the decomposition $H = \oplus_{i\in I} H_i$ is the array $(T_{ij})_{i,j\in I}$.

1. Prove that if $h = (h_i)_{i\in I}$, then $Th$ is given by multiplying the "column vector" $(h_i)_{i\in I}$ by the matrix $(T_{ij})_{i,j\in I}$, that is

$$Th = \left(\sum_{j\in I} T_{ij} h_j\right)_{i\in I}.$$

2. Prove that if $S \in B(H)$, then the block matrix decomposition of $ST$ with respect to the decomposition $H = \oplus_{i\in I} H_i$ is given by the ordinary rule for matrix multiplication, that is

$$(ST)_{ij} = \sum_k S_{ik} T_{kj},$$

where the above sum converges in the sense that for every $x \in H_j$, the series on the right-hand side of the following equation converges to the left-hand side:

$$(ST)_{ij}x = \sum_k S_{ik}T_{kj}x.$$

3. (Tricky) Construct an example of two operators $S$ and $T$ such that the series $\sum_k S_{ik}T_{kj}$ does not converge in norm.

4. Let $M$ be a closed subspace of Hilbert space $H$. Then every operator $T \in B(H)$ has a block decomposition

$$T = \begin{pmatrix} T_{11} & T_{12} \\ T_{21} & T_{22} \end{pmatrix}$$

with respect to the decomposition $H = M \oplus M^{\perp}$. Here $T_{11} \in B(M)$, $T_{12} \in B(M^{\perp}, M)$, etc. Prove that there exists a bijection between bounded operators on $H$ and $2 \times 2$ operator block matrices as above (there is also a bijection between operator block $n \times n$ matrices and operators on $H$ given a direct sum decomposition $H = M_1 \oplus \ldots \oplus M_n$).

5. Show that $TM \subseteq M$ if and only if $T$ has upper triangular form with respect to the decomposition $H = M \oplus M^{\perp}$ (that is if and only if $T_{21} = 0$).

6. Find the block decomposition of the orthogonal projection $P_M$. Prove that $TM \subseteq M$ if and only if $P_M T P_M = T P_M$, and use this to verify the previous item.

7. What is the block matrix decomposition of $T|_M$?

**Exercise 5.7.15.** Let $\{M_i\}_{i \in I}$ be a family of pairwise orthogonal closed subspaces of a Hilbert space $H$. Assume that the linear space spanned by sums of elements from the $M_i$s is dense in $H$ — in this case we say that $\{M_i\}_{i \in I}$ **spans** $H$ and we write $\oplus_{i \in I} M_i = H$. Suppose that we are given a family $\{A_i\}_{i \in I}$ of operators $A_i \in B(M_i)$, and that $\sup_i \|A_i\| < \infty$. Prove that there exists a unique operator $A = \oplus_{i \in I} A_i \in B(H)$, such that $Am = \oplus_{i \in I} A_i m = A_i m$ for all $i \in I$ and all $m \in M_i$. Show that for all $i$, $M_i$ is *reducing* for $A$ (meaning that $AM_i \subseteq M_i$ and $AM_i^{\perp} \subseteq M_i^{\perp}$). Calculate $\|\oplus_{i \in I} A_i\|$ and $(\oplus_{i \in I} A_i)^*$.

**Exercise 5.7.16** (Partial isometries). Let $H$ and $K$ be Hilbert spaces. A **partial isometry** from $H$ to $K$ is an operator $V \in B(H, K)$, such that the restriction $V|_{(\ker V)^{\perp}}$ of $V$ to the orthogonal complement of the kernel is an isometry. The space $(\ker V)^{\perp}$ is called the **initial space** of $V$, and $\operatorname{Im} V$ is called the **final space** of $V$. Prove that if $V$ is a partial isometry, then $\operatorname{Im} V$ is closed. Prove also that $V$ is a partial isometry if and only if one of the following conditions holds:

1. $V^*V$ is an orthogonal projection.

2. $V^*$ is a partial isometry.

3. $VV^*V = V$.

What is the relation between $V$ and the projections $V^*V$ and $VV^*$?

**Exercise 5.7.17.** Let $\{u_n\}_{n=1}^{\infty}$ be an orthonormal system of functions in $L^2[0,1]$. Let $L^2[-\pi,\pi]$ be given the normalized inner product $\langle f,g \rangle = \frac{1}{\pi}\int_{-\pi}^{\pi} f(t)\bar{g}(t)dt$.

1. Prove that there exists a unique bounded linear operator $T : L^2[-\pi,\pi] \to L^2[0,1]$ that satisfies
$$T(\cos nx) = 0 \quad \text{for all } n = 0,1,2,\ldots$$
   and
$$T(\sin nx) = u_n \quad \text{for all } n = 1,2,\ldots.$$

2. Prove that $T$ is a partial isometry. Determine the initial space of $T$.

3. Suppose that
$$\sum_{n=1}^{\infty}\left|\int_0^x u_n(t)dt\right|^2 = \min\left\{x,\frac{1}{3}\right\}.$$
   for all $x \in [0,1]$. Determine the final space of $T$.

**Exercise 5.7.18.** Fix $\lambda \in (0,1)$, and define an operator $T : C([0,1]) \to C([0,1])$ by
$$T(f)(t) = f(\lambda t).$$

1. Prove that $T$ is bounded, and thus extends to a bounded linear operator on $L^2[0,1]$ (the extension will also be denoted by $T$).

2. Find $\|T\|$, $T^*$, $T^*T$, $TT^*$ as well as the kernel and the image of $T$.

3. Find all continuous fixed vectors of $T$, that is, all $f \in C([0,1])$ such that $Tf = f$.

4. Is it possible for a function $f \in L^2[0,1]$ which is not in $C([0,1])$ to be a fixed vector of $T$, for some $\lambda \in (0,1)$? (Explaining this rigorously might be a little tricky; argue loosely.)

**Exercise 5.7.19.** Let $M$ and $N$ be two linear subspaces in a Hilbert space $H$ such that $M \cap N = \{0\}$. Let $M + N$ be the linear subspace defined by
$$M + N = \{m + n : m \in M, n \in N\},$$
and let $T : M + N \to K$ be a linear operator from $M + N$ to another Hilbert space $K$. Assume that the following conditions hold:

1. $\overline{M} \cap N = \{0\}$,

2. $N$ is finite dimensional,

3. $T|_M$ is bounded.

Prove that $T$ is bounded on $M + N$. Show that if any one of the above three conditions is violated, then one cannot conclude that $T$ is bounded.

**Exercise 5.7.20.** Let $\{T_n\}_{n=1}^{\infty}$ be a sequence of operators on a Hilbert space $H$ such that $\sup_n \|T_n\| \leq A$, and suppose that

$$T_i T_j^* = 0 \quad \text{for all } i \neq j,$$

and that

$$\|T_i^* T_j\| \leq c_i c_j \quad \text{for all } i \neq j,$$

where $\{c_i\}_{i=1}^{\infty}$ is a sequence of nonnegative numbers such that $\sum_{i=1}^{\infty} c_i = A$. Prove that the sum $\sum_{i=1}^{\infty} T_i$ determines a bounded operator $T$ on $H$ with $\|T\| \leq \sqrt{2}A$. (**Hint:** to be precise, what must be proved is that for every $h \in H$, the limit $Th = \lim_{n \to \infty} \sum_{i=1}^{n} T_i h$ exists and satisfies $\|Th\| \leq \sqrt{2}A\|h\|$; prove first the case where all $T_i = 0$ except for finitely many values of $i$, and then recall Exercise 2.5.2.)

# Chapter 6

## Hilbert function spaces

In this chapter, we present a couple of interesting applications of the Hilbert space theory we learned thus far, in the setting of *Hilbert function spaces*. The following chapters do not depend on this material, which is not essential for further study in functional analysis. The reader who is in a hurry to study the core of the basic theory may skip this chapter safely.

### 6.1 Hilbert function spaces

#### 6.1.1 The basic definition

**Definition 6.1.1.** A *Hilbert function space* on a set $X$ is a Hilbert space $H$ which is a linear subspace of the vector space $\mathbb{C}^X$ of all functions $X \to \mathbb{C}$, such that for every $x \in X$, the linear functional of point evaluation at $x$, given by $f \mapsto f(x)$, is a bounded functional on $H$.

The crucial property of point evaluations being bounded connects between the function theoretic properties of $H$ as a space of functions, and the Hilbert space structure of $H$, and therefore also the operator theory on $H$. Hilbert function spaces are also known as *reproducing kernel Hilbert spaces* (or *RKHS* for short), for reasons that will become clear very soon.

When we say that $H$ is a subspace of $\mathbb{C}^X$, we mean two things: (1) $H$ is a subset of $\mathbb{C}^X$; and (2) $H$ inherits its linear space structure from $\mathbb{C}^X$. In particular, the zero function is the zero element of $H$, and two elements of $H$ which are equal as functions are equal as elements of $H$. Moreover, we add functions and multiply them by scalars in the usual way.

**Example 6.1.2.** The Hilbert space $\ell^2 = \ell^2(\mathbb{N})$ is a space of functions from $\mathbb{N}$ to $\mathbb{C}$, and point evaluation is clearly bounded. Thus $\ell^2$ is a Hilbert function space on the set $\mathbb{N}$. However, it is not very interesting as a Hilbert function space, because the function theory of $\ell^2$ is not very rich.

In fact, every Hilbert space is a Hilbert function space on itself, by associating with every element $g$ the function $\phi_g : h \mapsto \langle g, h \rangle$ (why not $h \mapsto \langle h, g \rangle$?), but this is, again, not a very interesting example. Perhaps the most interesting examples of Hilbert function spaces are those that occur when $X$ is an open subset of $\mathbb{C}^d$, and now we shall meet one of the most beloved examples.

**Example 6.1.3** (The Hardy space). Let $\mathbb{D}$ denote the open unit disc in the complex plane $\mathbb{C}$. Recall that every function $f$ analytic in $\mathbb{D}$ has a power series representation $f(z) = \sum_{n=0}^{\infty} a_n z^n$. We define the **Hardy space** $H^2(\mathbb{D})$ to be the space of analytic functions on the disc with square summable Taylor coefficients, that is

$$H^2(\mathbb{D}) = \left\{ f(z) = \sum_{n=0}^{\infty} a_n z^n : \sum |a_n|^2 < \infty \right\}.$$

This space is endowed with the inner product

$$\left\langle \sum a_n z^n, \sum b_n z^n \right\rangle = \sum a_n \overline{b_n}. \tag{6.1}$$

Before worrying about function theoretic issues, we want to explain why $H^2(\mathbb{D})$ is a Hilbert space. Note that the space of formal power series with square summable coefficients, when endowed with the inner product (6.1), is obviously isomorphic (as an inner product space) to $\ell^2$. Thus to show that $H^2(\mathbb{D})$ is a Hilbert space, it suffices to show that it is the same as the space of all formal power series with square summable coefficients. Indeed, if $f(z) = \sum_{n=0}^{\infty} a_n z^n$ is a power series with square summable coefficients, then its coefficients are bounded, so it is absolutely convergent in the open unit disc. Moreover, for all $w \in \mathbb{D}$,

$$|f(w)| \leq \sum |a_n w^n| \leq \left( \sum |a_n|^2 \right)^{1/2} \left( \sum |w|^{2n} \right)^{1/2}. \tag{6.2}$$

Thus $H^2(\mathbb{D})$ is a space of analytic functions which as an inner product space is isomorphic to $\ell^2$, and is therefore complete. By (6.2), the space $H^2(\mathbb{D})$ is a Hilbert function space on $\mathbb{D}$, since for every $w \in \mathbb{D}$, the functional $f \mapsto f(w)$ is bounded by $(1 - |w|^2)^{-1/2}$.

By the Riesz representation theorem, this functional being bounded implies that there exists some $k_w \in H^2(\mathbb{D})$ such that $f(w) = \langle f, k_w \rangle$ for all $f \in H^2(\mathbb{D})$. Taking another look at (6.1), recalling that $f(w) = \sum a_n w^n$, and putting these two things together it is not hard to come up with the guess that $k_w(z) = \sum_{n=0}^{\infty} \overline{w}^n z^n$. For $|w| < 1$, we have $\sum |\overline{w}^n|^2 < \infty$, so $k_w \in H^2(\mathbb{D})$. Plugging $k_w$ in (6.1) we find

$$\langle f, k_w \rangle = \left\langle \sum a_n z^n, \sum \overline{w}^n z^n \right\rangle = \sum a_n w^n = f(w).$$

We will see below that there is a systematic way to find $k_w$.

## 6.1.2 Reproducing kernels

Let $H$ be a Hilbert function space on a set $X$. As point evaluation is a bounded functional, the Riesz representation theorem implies that there is, for every $x \in X$, an element $k_x \in H$ such that $f(x) = \langle f, k_x \rangle$ for all $f \in H$. The function $k_x$ is called the **kernel function** at $x$. We define a function on $X \times X$ by

$$k(x, y) = k_y(x).$$

The function $k(\cdot, \cdot)$ is called the **reproducing kernel** of $H$ or simply the **kernel** of $H$. The terminology comes from the fact that the functions $k_y = k(\cdot, y)$ enable one to reproduce the values of any $f \in H$ via the relationship $f(y) = \langle f, k_y \rangle$. Note that $k(x, y) = \langle k_y, k_x \rangle$.

**Example 6.1.4.** The kernel function for $H^2(\mathbb{D})$ is

$$k(z, w) = k_w(z) = \sum \overline{w}^n z^n = \frac{1}{1 - z\overline{w}}.$$

The kernel of the Hardy space is called the **Szegő kernel.**

**Exercise 6.1.5.** What is the kernel function of $\ell^2(\mathbb{N})$?

The kernel $k$ has the following important property: If $a_1, \ldots, a_n \in \mathbb{C}$ and $x_1, \ldots, x_n \in X$, then

$$\sum_{i,j=1}^n a_i \overline{a_j} k(x_j, x_i) = \sum_{i,j=1}^n a_i \overline{a_j} \langle k_{x_i}, k_{x_j} \rangle = \left\| \sum_{i=1}^n a_i k_{x_i} \right\|^2 \geq 0.$$

It follows that for every choice of points $x_1, \ldots, x_n \in X$, the matrix $\left( k(x_j, x_i) \right)_{i,j=1}^n$ is a positive semidefinite matrix, meaning that

$$\left\langle \left( k(x_j, x_i) \right)_{i,j=1}^n v, v \right\rangle \geq 0 \quad \text{for all } v \in \mathbb{C}^n.$$

A function with this property is called a **positive semidefinite kernel.** Every Hilbert function space gives rise to a positive semidefinite kernel — its reproducing kernel. In fact, there is a bijective correspondence between positive semidefinite kernels and Hilbert function spaces.

**Theorem 6.1.6.** *Let $k$ be a positive semidefinite kernel on $X$. Then there exists a unique Hilbert function space $H$ on $X$ such that $k$ is the reproducing kernel of $H$.*

**Remark 6.1.7.** In the context of Hilbert spaces, one might mistakenly understand that the uniqueness claimed is only up to some kind of Hilbert space isomorphism. To clarify, the uniqueness assertion is that there exists a unique subspace $H$ of the space $\mathbb{C}^X$ of all functions on $X$, which is a Hilbert function space, and has $k$ as a reproducing kernel.

*Proof.* Define $k_x = k(\cdot, x)$. Let $G_0$ be the linear subspace of $\mathbb{C}^X$ spanned by the functions $\{k_x : x \in X\}$. Equip $G_0$ with the sesquilinear form

$$\left\langle \sum a_i k_{x_i}, \sum b_j k_{y_j} \right\rangle_0 = \sum a_i \overline{b_j} k(y_j, x_i).$$

From the definition and the fact that $k$ is a positive semidefinite kernel, the form $\langle \cdot, \cdot \rangle_0$ satisfies all the properties of an inner product, except, perhaps, that it is not definite; that is, the only property missing is $\langle g, g \rangle_0 = 0 \Rightarrow g = 0$. This means that the Cauchy-Schwartz inequality holds for this form on $G_0$ (recall Exercise 2.2.9). From this, it follows that the space $N = \{g \in G_0 : \langle g, g \rangle_0 = 0\}$ is a subspace, and also that $\langle f, g \rangle_0 = 0$ for all $f \in G_0$ and $g \in N$. This implies that the quotient space $G = G_0/N$ can be equipped with the inner product

$$\langle \dot{f}, \dot{g} \rangle = \langle f, g \rangle_0.$$

Here we are using the notation $\dot{f}$ to denote the equivalence class of $f$ in $G_0/N$. That this is well-defined and an inner product follows from what we said about $N$. Now, complete $G$, using Theorem 2.3.1, to obtain a Hilbert space $H$.

We started with a space of functions on $X$, but by taking the quotient and completing we are now no longer dealing with functions on $X$. To fix this, we define a map $F : H \to \mathbb{C}^X$ by

$$F(h)(x) = \langle h, \dot{k}_x \rangle. \tag{6.3}$$

This is a linear map and it is injective because the set $\{\dot{k}_x : x \in X\}$ spans a dense subspace of $G$. Note that

$$F(\dot{k}_x)(y) = \langle \dot{k}_x, \dot{k}_y \rangle = k(y, x). \tag{6.4}$$

We may push the inner product from $H$ to $F(H)$, that is, we define $\langle F(h), F(g) \rangle$ by

$$\langle F(h), F(g) \rangle = \langle h, g \rangle,$$

and we obtain a Hilbert space of functions in $\mathbb{C}^X$. Point evaluation is bounded by definition (see (6.3)), and the kernel function at $x$ is $F(\dot{k}_x)$. By (6.4), $F(\dot{k}_x) = k_x$.

Let us now identify between $H$ and $F(H)$. It follows that $H$ is spanned by the kernel functions $\{k_x : x \in X\}$. The kernel function of this Hilbert function space $H$ is given by

$$\langle \dot{k}_y, \dot{k}_x \rangle = \langle k_y, k_x \rangle = k(x, y),$$

as required.

Now suppose that $L$ is another Hilbert function space on $X$ that has $k$ as a reproducing kernel. By definition, the kernel functions are contained in $L$, and since $H$ is the closure of the space spanned by the kernel functions, $H \subseteq L$. However, if $f$ is in the orthogonal complement of $H$ in $L$, then $f(x) = \langle f, k_x \rangle = 0$ for all $x \in X$, thus $f \equiv 0$. This shows that $L = H$, completing the proof. □

**Exercise 6.1.8.** Let $H$ be a RKHS with kernel $k$. Let $x_1, \ldots, x_n \in X$. Show that the matrix $\big(k(x_i, x_j)\big)_{i,j=1}^n$ is positive definite (meaning that it is positive semidefinite and invertible) if and only if the functions $k_{x_1}, \ldots, k_{x_n}$ are linearly independent, and this happens if and only if the evaluation functionals at the points $x_1, \ldots, x_n$ are linearly independent as functionals on $H$.

**Exercise 6.1.9.** Show how the proof of the above theorem is simplified when $k$ is assumed strictly positive definite (meaning that the matrix $\left(k(x_i, x_j)\right)_{i,j=1}^{n}$ is strictly positive definite for any choice of points). Show how at every stage of the process one has a space of functions on $X$.

**Proposition 6.1.10.** *Let $H$ be a Hilbert function space on a set $X$. Let $\{e_i\}_{i \in I}$ be an orthonormal basis for $H$. Then the kernel $k$ for $H$ is given by the formula*

$$k(x, y) = \sum_{i \in I} e_i(x)\overline{e_i(y)}, \tag{6.5}$$

*where the sum converges (absolutely) at every $x, y \in X$.*

*Proof.* We use the generalized Parseval identity

$$
\begin{aligned}
k(x, y) &= \langle k_y, k_x \rangle \\
&= \sum_{i \in I} \langle k_y, e_i \rangle \langle e_i, k_x \rangle \\
&= \sum_{i \in I} e_i(x)\overline{e_i(y)}.
\end{aligned}
$$

$\square$

**Example 6.1.11** (The Hardy space, revisited). We can now see how to come up with the formula for the Szegő kernel (the kernel for the Hardy space). An orthonormal basis for $H^2(\mathbb{D})$ is given by $1, z, z^2, \ldots$, so by the proposition

$$k(z, w) = \sum_{n=0}^{\infty} z^n \overline{w^n} = \frac{1}{1 - z\bar{w}}.$$

---

## 6.2   *The Bergman space

### 6.2.1   Some basic facts on analytic functions

In this section, we will require some notations and some very basic facts about analytic functions. To keep the treatment self-contained, we will use almost only facts that follow from first-year undergraduate material on power series, and we will work with a definition of *analytic function* which is not the standard definition (it turns out that this definition is equivalent to the usual one). One fact that is more advanced and that we shall need is the fact that the uniform limit of analytic functions is analytic. This fact — which appears in any course on complex analysis — will follow from our results in the next section; see Remark 6.3.11 (see [30] for the standard proof).

For $z_0 \in \mathbb{C}$ and $r > 0$, we let $D_r(z_0)$ denote the disc with center $z_0$ and radius $r$, that is

$$D_r(z_0) = \{z \in \mathbb{C} : |z_0 - z| < r\}.$$

**Definition 6.2.1.** A function $f : D_r(z_0) \to \mathbb{C}$ is said to be **analytic** (or **analytic in** $D_r(z_0)$) if it has a power series representation

$$f(z) = \sum_{n=0}^{\infty} a_n (z - z_0)^n$$

that converges for every $z \in D_r(z_0)$.

The space of analytic functions on a disc $D$ is denoted by $\mathcal{O}(D)$. The following lemma is a well-known consequence of the theory of functions of a complex variable, but it can also be proved directly using power series.

**Lemma 6.2.2.** *Suppose that $D$ is open disc. If $f \in \mathcal{O}(D)$, and $z_0 \in D$, then $f$ is analytic in $D_r(z_0)$ as well, for any $r > 0$ such that $D_r(z_0) \subseteq D$.*

In other words, if a function has a power series representation around a point in a disc, then it has a power series representation (converging in a smaller disc, perhaps) around any other point of that disc. We leave the proof, the acceptance in blind faith, or the task of looking up a proof[1] to the reader.

From now on, we identify $\mathbb{C}$ with $\mathbb{R}^2$, and consider a function of the complex variable $z = x + iy$ to be a function of two real variables $x, y$.

**Lemma 6.2.3.** *If $D$ is an open disc, $f \in \mathcal{O}(D)$, and $\overline{D_r(z_0)} \subset D$, then*

$$f(z_0) = \frac{1}{\pi r^2} \int_{D_r(z_0)} f(x + iy) dx dy. \tag{6.6}$$

In other words, the value of $f$ at $z_0$ is equal to the mean of $f$ in a disc around $z_0$.

*Proof.* By translating the variable, we may assume that $z_0 = 0$. So we can write $f(z) = \sum_{n=0}^{\infty} a_n z^n$, where this sum converges uniformly in $D_r(0)$. The integral can therefore be evaluated term by term. A direct calculation using polar coordinates shows that $\int_{D_r(0)} (x + iy)^n dx dy = 0$ for all $n \geq 1$, so

$$\frac{1}{\pi r^2} \int_{D_r(0)} f(x + iy) dx dy = \frac{1}{\pi r^2} \int_{D_r(0)} a_0 dx dy = a_0 = f(0),$$

as required.                                                                        □

---

[1] See [28, Theorem 8.4] for an elementary proof, or read [30] to understand the result in the context of complex function theory.

## 6.2.2 Definition of the Bergman space

Recall that $\mathbb{D}$ denotes the unit disc $\mathbb{D} = \{z \in \mathbb{C} : |z| < 1\}$. The **Bergman space** $L_a^2(\mathbb{D})$ is the space of all analytic functions in the disc which are square integrable on the disc with respect to Lebesgue measure, i.e.,

$$L_a^2(\mathbb{D}) = L^2(\overline{\mathbb{D}}) \cap \mathcal{O}(\mathbb{D}).$$

Of course no measure theory is required for this definition: analytic functions are smooth, so they are integrable on every smaller disc, and for every $f \in \mathcal{O}(\mathbb{D})$ one may define

$$\|f\|_2 = \lim_{r \to 1} \sqrt{\int_{x^2+y^2<r} |f(x+iy)|^2 dx dy} \tag{6.7}$$

and then

$$L_a^2(\mathbb{D}) = \{f \in \mathcal{O}(\mathbb{D}) : \|f\|_2 < \infty\}.$$

This space has an inner product defined by

$$\langle f, g \rangle = \int_{\mathbb{D}} f\bar{g}\, dx dy = \lim_{r \to 1} \int_{x^2+y^2<r} f\bar{g}\, dx dy.$$

**Lemma 6.2.4.** *For every $z_0 \in \mathbb{D}$, the linear functional $f \mapsto f(z_0)$ is bounded on $L_a^2(\mathbb{D})$. In fact, if $|z_0| < 1 - r$ then*

$$|f(z_0)| \le \frac{1}{\sqrt{\pi r}} \|f\|_2. \tag{6.8}$$

*Proof.* Let $r > 0$ be such that $r < 1 - |z_0|$; equivalently, $\overline{D_r(z_0)} \subset \mathbb{D}$. By (6.6) we find that

$$
\begin{aligned}
|f(z_0)| &= \frac{1}{\pi r^2} \left| \int_{D_r(z_0)} f \right| \\
&\le \frac{1}{\pi r^2} \left( \int_{D_r(z_0)} |f|^2 \right)^{1/2} \left( \int_{D_r(z_0)} 1 \right)^{1/2} \\
&\le \frac{1}{\pi r^2} \left( \int_{\mathbb{D}} |f|^2 \right)^{1/2} \sqrt{\pi} r \\
&= \frac{1}{\sqrt{\pi r}} \|f\|_2,
\end{aligned}
$$

where the first inequality is the Cauchy-Schwarz inequality for integrals, and the second inequality follows from monotonicity of the integral (integrating a positive function on a bigger set gives a bigger value). $\qquad\square$

### 6.2.3   The Bergman space as a Hilbert function space

**Theorem 6.2.5.** *The Bergman space is a Hilbert function space.*

*Proof.* That point evaluation is bounded follows from (6.8). It remains to show completeness.

Let $\{f_n\}$ be a Cauchy sequence in $L_a^2(\mathbb{D})$. By (6.8) once more, we find that for every $z_0$ in the disc $D_{1-r}(0)$ we have

$$|f_n(z_0) - f_m(z_0)| \le \frac{1}{\sqrt{\pi}r}\|f_n - f_m\|_2. \tag{6.9}$$

It follows that $\{f_n\}$ is a Cauchy sequence also in the uniform norm on $\mathcal{O}(D_{1-r}(0))$. Since this is true for every $r > 0$, we find that $\{f_n\}$ is pointwise convergent in $\mathbb{D}$, say $f_n(z) \to f(z)$ for all $z \in \mathbb{D}$.

Now $\{f_n\}$ converges uniformly in $D_{1-r}(0)$, and since the uniform limit of analytic functions is analytic (see Section VII.15 in [30]), we have that $f$ is analytic in $D_{1-r}(0)$. Since this is true for every $r > 0$, we have $f \in \mathcal{O}(\mathbb{D})$. Since $L^2(\overline{\mathbb{D}})$ is complete, $f_n$ converges to some element in $L^2(\overline{\mathbb{D}})$, and this element must be $f$; thus $f \in L_a^2(\mathbb{D})$, and the proof is complete.

Did the last line of the proof convince the reader? It shouldn't have, if the student did not take a course on measure theory. Let us give another explanation, with a few more details, regarding the last step. We need to show that $\|f\|_2 < \infty$ and that $\|f_n - f\|_2 \to 0$. Fix $\epsilon > 0$, and let $N \in \mathbb{N}$ such that for all $m, n \ge N$, $\|f_n - f_m\|_2 < \epsilon$. For every $\rho \in (0, 1)$

$$\sqrt{\int_{x^2+y^2<\rho} |f_n - f_m|^2} \le \|f_n - f_m\|_2 < \epsilon.$$

Since $f_m \to f$ uniformly in $D_\rho(0)$, we take the limit as $m \to \infty$ above to obtain

$$\sqrt{\int_{x^2+y^2<\rho} |f_n - f|^2} \le \epsilon,$$

for all $n \ge N$. As this holds for every $\rho \in (0, 1)$, we get that $f_n - f \in L_a^2(\mathbb{D})$, thus $f \in L_a^2(\mathbb{D})$. Moreover we obtain that $\|f_n - f\|_2 \le \epsilon$, for all $n \ge N$, therefore $f_n \to f$ in $L^2$. This completes the proof. $\qquad\square$

Since $L_a^2(\mathbb{D})$ is a reproducing kernel Hilbert space, we know that it has a reproducing kernel. We now compute the kernel.

**Proposition 6.2.6.** *The kernel of $L_a^2(\mathbb{D})$ is given by*

$$k(z, w) = \frac{1}{\pi}\frac{1}{(1 - z\bar{w})^2}.$$

*Thus, for every $f \in L_a^2(\mathbb{D})$ and every $z_0 = x_0 + iy_0 \in \mathbb{D}$, we have the reproducing formula*

$$f(z_0) = \frac{1}{\pi}\int_{\mathbb{D}} \frac{f(x + iy)dxdy}{(1 - (x + iy)(x_0 - iy_0))^2}. \tag{6.10}$$

*Proof.* We wish to use Proposition 6.1.10. For that, we need to find first an orthonormal basis for $L_a^2(\mathbb{D})$. The sequence of monomials $e_n(z) = z^n$, (for $n \geq 0$) is an obvious candidate. We use the polar representation $z = re^{i\theta}$, and calculate

$$\langle z^m, z^n \rangle = \int_0^1 \int_0^{2\pi} r^{m+n} e^{i\theta(m-n)} r\, d\theta dr$$

and this 0 if $m \neq n$ and equal to $\pi/(n+1)$ when $m = n$. An orthonormal basis is therefore given by

$$\frac{1}{\sqrt{\pi}}, \sqrt{\frac{2}{\pi}} z, \sqrt{\frac{3}{\pi}} z^2, \ldots, \sqrt{\frac{n+1}{\pi}} z^n, \ldots$$

(The completeness of this sequence in $L_a^2(\mathbb{D})$ is easy to establish). Using Proposition 6.1.10 we find that the kernel of the Bergman space is given by

$$\sum_{n=0}^{\infty} \frac{n+1}{\pi} (z\bar{w})^n = \frac{1}{\pi} \frac{1}{(1 - z\bar{w})^2}.$$

Now (6.10) follows from the reproducing property of the kernel. $\qquad\square$

---

## 6.3  *Additional topics in Hilbert function space theory

In this section, we will occasionally use a result or a notion from a later chapter. Most readers will be able to read and enjoy it now; those who insist on learning everything in a linear order may return to it after Chapter 8.

### 6.3.1  $H^2(\mathbb{D})$ as a subspace of $L^2(\mathbb{T})$

We now continue to study the space $H^2(\mathbb{D})$, and give a useful formula for the norm of elements in this space.

Let $f(z) = \sum_{n=0}^{\infty} a_n z^n \in H^2(\mathbb{D})$. For any $r \in (0,1)$, define $f_r(z) = f(rz)$. The power series of $f_r$ is given by

$$f_r(z) = \sum_{n=0}^{\infty} a_n r^n z^n.$$

This series converges to $f_r$ uniformly on the closed unit disc. In fact, $f_r$ extends to an analytic function on an open neighborhood of the closed unit disc, and the series converges uniformly and absolutely on a closed disc that is slightly larger than the unit disc. It follows that the following computations are valid:

$$\frac{1}{2\pi}\int_0^{2\pi}|f_r(e^{it})|^2dt = \sum_{m,n=0}^{\infty}\frac{1}{2\pi}\int_0^{2\pi}a_mr^me^{imt}\overline{a_n}r^ne^{-int}dt$$

$$= \sum_{n=0}^{\infty}r^{2n}|a_n|^2 = \|f_r\|_{H^2}^2.$$

It would be nice to say that for every $f \in H^2(\mathbb{D})$

$$\|f\|_{H^2} = \left(\frac{1}{2\pi}\int_0^{2\pi}|f(e^{it})|^2dt\right)^{1/2},$$

and this is almost true, but we have to be careful because elements in $H^2(\mathbb{D})$ are not defined on $\mathbb{T} = \{z \in \mathbb{C} : |z| = 1\}$. However, we do have the following characterization of the norm of $H^2(\mathbb{D})$.

**Lemma 6.3.1.** *A function $f \in \mathcal{O}(\mathbb{D})$ is in $H^2(\mathbb{D})$ if and only if*

$$\lim_{r\to1}\int_0^{2\pi}|f(re^{it})|^2dt < \infty.$$

*If $f \in H^2(\mathbb{D})$, then*

$$\|f\|^2 = \lim_{r\to1}\frac{1}{2\pi}\int_0^{2\pi}|f(re^{it})|^2dt.$$

*Proof.* Let $f$ be analytic in the disc, and suppose it has power series representation $f(z) = \sum_{n=0}^{\infty}a_nz^n$. As before

$$\frac{1}{2\pi}\int_0^{2\pi}|f(re^{it})|^2 = \sum_{n=0}^{\infty}r^{2n}|a_n|^2.$$

The right-hand side converges to $\sum_{n=0}^{\infty}|a_n|^2$ (justify this), so the limit of the integrals is finite if and only if $f \in H^2(\mathbb{D})$. $\qquad\square$

**Remark 6.3.2.** It can be shown that much more is true: as $r \nearrow 1$, the functions $e^{it} \mapsto f_r(e^{it})$ converge almost everywhere to a function $f^* \in L^2(\mathbb{T})$, and $\|f\|_{H^2} = \|f^*\|_{L^2}$. Moreover, $H^2(\mathbb{D})$ can be identified via the map $f \mapsto f^*$ with the subspace of $L^2(\mathbb{T})$ consisting of all functions with Fourier series that are supported on the nonnegative integers: if $f(z) = \sum_{n=0}^{\infty}a_nz^n \in H^2(\mathbb{D})$, then $f^*(e^{it}) = \sum_{n=0}^{\infty}a_ne^{int}$. For details, see [2].

### 6.3.2 Multiplier algebras

Let $H$ be a Hilbert function space on $X$, to be fixed in the following discussion. We now define a class of natural operators on $H$.

**Definition 6.3.3.** The *multiplier algebra of H*, Mult($H$), is defined to be

$$\text{Mult}(H) = \{f : X \to \mathbb{C} : fh \in H \text{ for all } h \in H\}.$$

An element $f \in \text{Mult}(H)$ is called a *multiplier*.

In this definition $fh$ denotes the usual pointwise product of two functions: $[fh](x) = f(x)h(x)$. For every $f \in \text{Mult}(H)$ we define an operator $M_f : H \to H$ by

$$M_f h = fh.$$

It is not unusual to identify $M_f$ and $f$, and we will do so too, but sometimes it is helpful to distinguish between the function $f$ and the linear operator $M_f$.

**Proposition 6.3.4.** *For all $f \in \text{Mult}(H)$, we have $M_f \in B(H)$.*

*Proof.* Obviously $M_f$ is a well-defined linear operator. To show that it is bounded, we use the closed graph theorem (see Exercise 8.5.3). Let $h_n \to h \in H$, and suppose that $M_f h_n = fh_n \to g \in H$. Since point evaluation is continuous we have for all $x \in X$

$$h_n(x) \to h(x)$$

and

$$f(x)h_n(x) \to g(x),$$

thus $fh(x) = g(x)$ for all $x$, meaning that $M_f h = g$, so the graph of $M_f$ is closed. By the closed graph theorem, we conclude that $M_f$ is bounded. $\square$

Thus, $\{M_f : f \in \text{Mult}(H)\}$ is a subalgebra of $B(H)$. We norm Mult($H$) by pulling back the norm from $B(H)$ (this means that we define $\|f\|_{\text{Mult}(H)} = \|M_f\|$), and this turns Mult($H$) into a normed algebra. We will see below that it is a Banach algebra, under a natural assumption.

**Standing assumption.** *From now on, we assume that for every $x \in X$ there exists a function $h \in H$ such that $h(x) \neq 0$.*

This assumption holds in most (but not all) cases of interest, and in particular in the case of $H^2(\mathbb{D})$. It is equivalent to the assumption that $k_x(x) = k(x,x) = \|k_x\|^2 \neq 0$ for all $x \in X$. In fact, when the assumption does not hold, then $\|\cdot\|_{\text{Mult}(H)}$ is not a norm, but only a seminorm.

One of the instances when it is helpful to make a distinction between a multiplier $f$ and the multiplication operator $M_f$ that it induces is when discussing adjoints.

**Proposition 6.3.5.** *Let $f \in \text{Mult}(H)$. For all $x \in X$, the kernel function $k_x$ is an eigenvector for $M_f^*$ with eigenvalue $\overline{f(x)}$ (see Definition 8.3.16). Conversely, if $T \in B(H)$ is an operator such that $Tk_x = \lambda(x)k_x$ for all $x \in X$, then there is an $f \in \text{Mult}(H)$ such that $\lambda(x) = \overline{f(x)}$ for all $x$, and $T = M_f^*$.*

*Proof.* We let $x \in X$ and compute for all $h \in H$

$$f(x)h(x) = \langle M_f h, k_x \rangle = \langle h, M_f^* k_x \rangle.$$

But $f(x)h(x) = \langle h, \overline{f(x)}k_x \rangle$. Thus $M_f^* k_x = \overline{f(x)}k_x$.

For the converse, suppose $Tk_x = \lambda(x)k_x$ for all $x \in X$. Define $f(x) = \overline{\lambda(x)}$. For all $h \in H$, we have $T^* h \in H$ too, so

$$\langle T^* h, k_x \rangle = \langle h, Tk_x \rangle = \overline{\lambda(x)}\langle h, k_x \rangle = f(x)h(x).$$

It follows that $T^* h = fh \in H$, so $f \in \text{Mult}(H)$. It also follows that $T^* = M_f$.                                                                                       □

**Corollary 6.3.6.** $\text{Mult}(H)$ *contains only bounded functions, and*

$$\sup_X |f| \le \|M_f\|.$$

*Proof.* This follows from the Propositions 6.3.4 and 6.3.5, since the eigenvalues of $M_f^*$ are clearly bounded by $\|M_f^*\| = \|M_f\|$.                                        □

**Corollary 6.3.7.** $\{M_f : f \in \text{Mult}(H)\}$ *is closed in the norm of* $B(H)$. *In fact, if* $\{T_n\}$ *is a sequence in* $\{M_f : f \in \text{Mult}(H)\}$ *and* $T \in B(H)$ *satisfies*

$$\langle T_n f, g \rangle \xrightarrow{n \to \infty} \langle Tf, g \rangle \quad \text{for all } f, g \in H,$$

*then* $T \in \{M_f : f \in \text{Mult}(H)\}$.

**Remark 6.3.8.** If $\langle T_n f, g \rangle \to \langle Tf, g \rangle$ for all $f, g \in H$, then the sequence $\{T_n\}$ is said to converge to $T$ in the **weak operator topology**. This notion indeed corresponds to convergence of sequence in the so-called weak operator topology, one of several different useful topologies one can define on $B(H)$. In essence, we are proving here that $\{M_f : f \in \text{Mult}(H)\}$ is closed in the weak operator topology.

*Proof.* We will prove the second assertion, which immediately implies the first (why?). If $T_n = M_{f_n}$ is a sequence in $\{M_f : f \in \text{Mult}(H)\}$ converging to $T$ in the weak operator topology, then it follows that $T_n^* \to T^*$ in the weak operator topology. Then for all $x \in X$ and all $h \in H$,

$$f_n(x)h(x) = f_n(x)\langle h, k_x \rangle = \langle h, T_n^* k_x \rangle \to \langle h, T^* k_x \rangle.$$

By our standing assumption, the sequence $\{f_n\}$ converges pointwise, say to a limit $f_0$. We find that $\langle h, T^* k_x \rangle = f_0(x)h(x)$, so $k_x$ is an eigenvector for $T^*$ with an eigenvalue $\overline{f_0(x)}$. By Proposition 6.3.5, $T = M_{f_0} \in \{M_f : f \in \text{Mult}(H)\}$.                                                                    □

**Example 6.3.9.** Let us find out what is the multiplier algebra of $H^2(\mathbb{D})$. Since $1 \in H^2(\mathbb{D})$, we have that $\mathrm{Mult}(H) \subseteq H^2$, and in particular every multiplier in $\mathrm{Mult}(H)$ is an analytic function on $\mathbb{D}$. By Corollary 6.3.6 we have that every multiplier is bounded on the disc. Thus

$$\mathrm{Mult}(H^2(\mathbb{D})) \subseteq H^\infty(\mathbb{D}),$$

where $H^\infty(\mathbb{D})$ denotes the algebra of bounded analytic functions in $\mathbb{D}$. On the other hand, by the formula developed in Section 6.3.1, we have for $f \in H^\infty(\mathbb{D})$ and $h \in H^2(\mathbb{D})$,

$$\lim_{r \to 1} \frac{1}{2\pi} \int_0^{2\pi} |f(re^{it})h(re^{it})|^2 dt \le \|f\|_\infty^2 \lim_{r \to 1} \frac{1}{2\pi} \int_0^{2\pi} |h(re^{it})|^2 dt$$
$$= \|f\|_\infty^2 \|h\|_{H^2}^2.$$

By Lemma 6.3.1, we obtain $fh \in H^2(\mathbb{D})$ and $\|fh\|_{H^2} \le \|f\|_\infty \|h\|_{H^2}$, so $f \in \mathrm{Mult}(H^2(\mathbb{D}))$ and $\|M_f\| \le \|f\|_\infty$. We conclude that $\mathrm{Mult}(H^2(\mathbb{D})) = H^\infty(\mathbb{D})$, with the norm $\|f\|_{\mathrm{Mult}(H)} = \|f\|_\infty$.

**Exercise 6.3.10.** The *Segal-Bargmann space* is the space of all entire functions $f$ for which $\int_\mathbb{C} |f(x+iy)|^2 e^{-x^2-y^2} dxdy < \infty$. Show that this space is a Hilbert function space. What is the multiplier algebra of this space?

**Remark 6.3.11.** Note that as a consequence of Corollary 6.3.7 and Example 6.3.9, we have that $H^\infty(\mathbb{D})$ is a complete space. It follows readily that the limit of a uniformly convergent sequence of bounded analytic functions on any disc is also bounded and analytic in that disc. This, in turn, implies that if $\{f_n\}$ is a sequence of analytic functions on a domain $D \subset \mathbb{C}$, and if $f_n \to f$ uniformly on all compact subsets of $D$, then $f$ must be analytic in $D$. This is a nontrivial theorem in complex analysis, that is usually proved using the theory of complex integration. How much complex function theory did we use to obtain this result?

## 6.3.3   Pick's interpolation theorem

Given $n$ distinct points $z_1, \ldots, z_n$ in the unit disc, and $w_1, \ldots, w_n$ complex numbers, one can always find a polynomial $p$ of degree at most $n-1$ that satisfies

$$p(z_i) = w_i \quad \text{for all } i = 1, \ldots, n.$$

One then says that $p$ *interpolates* the data points $(z_1, w_1), \ldots, (z_n, w_n)$, and $p$ is said to be the *interpolating polynomial*. The existence and uniqueness of the interpolating polynomial is an exercise in linear algebra. It is nice to know that we can always map the points $z_i$ to the points $w_i$ with a function from within a well-understood class of functions, or, to say it the other way around, that we can fit the data points in the graph of a nice function. However, the interpolating polynomial may behave in unexpected ways.

**Exercise 6.3.12.** Fix $n \geq 2$. Given any $M > 0$, show that $z_1, \ldots, z_n \in \mathbb{D}$ and $w_1, \ldots, w_n \in \mathbb{C}$ can be chosen so that if $p$ is the interpolating polynomial of degree at most $n - 1$, then

$$\frac{\|p\|_\infty}{\max_i |w_i|} > M.$$

One is therefore led to consider the problem of interpolating the data with other kinds of nice functions, and removing the restriction on the degree of polynomials leads us quickly to analytic functions. The **Pick interpolation problem** is the problem of finding a bounded analytic function that interpolates given data points. In this section, we will show how the framework of Hilbert function spaces allows us to elegantly solve the Pick interpolation problem. The material in this section is adapted from Chapter 6 of the monograph [2], which contains an in-depth treatment of Pick interpolation.

**Theorem 6.3.13** (Pick's interpolation theorem). *Let* $z_1, \ldots, z_n \in \mathbb{D}$, *and* $w_1, \ldots, w_n \in \mathbb{C}$ *be given. There exists a function* $f \in H^\infty(\mathbb{D})$ *satisfying* $\|f\|_\infty \leq 1$ *and*

$$f(z_i) = w_i \quad \text{for all } i = 1, \ldots, n,$$

*if and only if the following matrix inequality holds:*

$$\left( \frac{1 - w_i \overline{w_j}}{1 - z_i \overline{z_j}} \right)^n_{i,j=1} \geq 0.$$

The proof of the theorem will occupy the remainder of this section. The necessary and sufficient condition for interpolation is that a certain $n \times n$ matrix, with the element in the $ij$-th place equal to $\frac{1 - w_i \overline{w_j}}{1 - z_i \overline{z_j}}$, is positive semidefinite. Recall that an $n \times n$ matrix $A$ is said to be **positive semidefinite** if the operator $T_A : x \mapsto Ax$ is a positive operator in the sense of Definition 5.5.1. If $A$ is positive semidefinite, then we write $A \geq 0$.

Note that the matrix element $\frac{1 - w_i \overline{w_j}}{1 - z_i \overline{z_j}}$ appearing in the theorem is equal to $(1 - w_i \overline{w_j})k(z_i, z_j)$, where $k(z, w) = \frac{1}{1 - z\overline{w}}$ is the Szegő kernel, that is, the reproducing kernel for the Hardy space $H^2(\mathbb{D})$. Given $z_1, \ldots, z_n \in \mathbb{D}$ and $w_1, \ldots, w_n \in \mathbb{C}$, the matrix

$$((1 - w_i \overline{w_j})k(z_i, z_j))^n_{i,j=1}$$

is referred to as the **Pick matrix**.

**A necessary condition for interpolation**

**Lemma 6.3.14.** *Let* $T \in B(H)$. *Then* $\|T\| \leq 1$ *if and only if* $I - T^*T \geq 0$.

*Proof.* Let $h \in H$. Then

$$\langle h, h \rangle - \langle Th, Th \rangle = \langle (I - T^*T)h, h \rangle \geq 0$$

is equivalent to

$$\|Th\|^2 \le \|h\|^2.$$

□

**Proposition 6.3.15.** *Let $H$ be a Hilbert function space on a set $X$ with a reproducing kernel $k$. A function $f : X \to \mathbb{C}$ is a multiplier of norm $\|f\|_{\mathrm{Mult}(H)} \le 1$, if and only if for every $n \in \mathbb{N}$, and every $n$ points $x_1, \ldots, x_n \in X$, the associated Pick matrix is positive semidefinite, that is:*

$$\left( (1 - f(x_i)\overline{f(x_j)})k(x_i, x_j) \right) \ge 0.$$

*Proof.* Let us define the operator $T : \mathrm{span}\{k_x : x \in X\} \to H$ by extending linearly the rule

$$Tk_x = \overline{f(x)}k_x.$$

Suppose first that $f$ is a multiplier of norm $\|f\|_{\mathrm{Mult}(H)} \le 1$. Then by Proposition 6.3.5, $T$ extends to $H$ and is equal to $T = M_f^*$. Let $x_1, \ldots, x_n \in X$. Then for any $c_1, \ldots, c_n \in \mathbb{C}$,

$$0 \le \left\langle (I - T^*T) \sum c_j k_{x_j}, \sum c_i k_{x_i} \right\rangle = \sum_{i,j} c_j \overline{c_i}(1 - f(x_i)\overline{f(x_j)})\langle k_{x_j}, k_{x_i} \rangle$$

$$= \sum_{i,j} c_j \overline{c_i}(1 - f(x_i)\overline{f(x_j)})k(x_i, x_j).$$

This implies that the Pick matrix is positive semidefinite.

Conversely, if the Pick matrix is positive semidefinite, the above calculation combined with the lemma shows that $T$ is a bounded operator on $\mathrm{span}\{k_x : x \in X\}$ with norm at most 1. Since the span of $k_x$s is dense in $H$, we conclude that $T$ extends to a contraction on $H$, and by Proposition 6.3.5 again, we obtain that $f$ is a multiplier and that $T = M_f^*$. □

**Corollary 6.3.16.** *Let $z_1, \ldots, z_n \in \mathbb{D}$, and $w_1, \ldots, w_n \in \mathbb{C}$ be given. A necessary condition for there to exist a function $f \in H^\infty(\mathbb{D})$ satisfying $\|f\|_\infty \le 1$ and*

$$f(z_i) = w_i \quad \text{for all } i = 1, \ldots, n,$$

*is that the following matrix inequality holds:*

$$\left( \frac{1 - w_i \overline{w_j}}{1 - z_i \overline{z_j}} \right)_{i,j=1}^n \ge 0.$$

*Proof.* This follows from the Proposition applied to the Hardy space, since $\mathrm{Mult}(H^2(\mathbb{D})) = H^\infty(\mathbb{D})$. □

## The realization formula

Let $H_1^\infty$ denote the closed unit ball of $H^\infty(\mathbb{D})$. A key to the solution of the interpolation problem is the following characterization of functions in $H_1^\infty$. The reader should recall the notion of operator block matrices (see Exercise 5.7.14).

**Theorem 6.3.17.** *Let $f : \mathbb{D} \to \mathbb{C}$. Then $f \in H_1^\infty$ if and only if there is a Hilbert space $K$ and an isometry $V : \mathbb{C} \oplus K \to \mathbb{C} \oplus K$ with block structure*

$$V = \begin{pmatrix} a & B \\ C & D \end{pmatrix},$$

*such that*

$$f(z) = a + zB(I - zD)^{-1}C \tag{6.11}$$

*for all $z \in \mathbb{D}$.*

*Proof.* For the proof of Pick's interpolation theorem, we shall only require that the condition that $f$ is given by (6.11) for some $V$ as in the theorem, implies that $f \in H_1^\infty$. Therefore, we will prove that this condition is sufficient, and leave the necessity of the condition as an exercise (see Exercise 6.3.21 for a hint).

For sufficiency, suppose that $V$ is an isometry with a block decomposition as in the theorem, and that $f$ is given by (6.11). Then $\|D\| \leq 1$, thus $I - zD$ is invertible for all $z \in \mathbb{D}$, and $(I - zD)^{-1}$ is an analytic operator valued function. To be precise, by Proposition 8.3.9 below, $(I - zD)^{-1}$ has a power series representation

$$(I - zD)^{-1} = \sum_{n=0}^\infty z^n D^n.$$

Therefore $f$ defined by (6.11) is analytic in $\mathbb{D}$, because

$$f(z) = a + \sum_{n=0}^\infty (BD^n C)z^{n+1}.$$

Note here that $B \in B(K, \mathbb{C})$, $D^n \in B(K)$ and $C \in B(\mathbb{C}, K)$, therefore $BD^n C \in B(\mathbb{C}) = \mathbb{C}$. In fact, $|BD^n C| \leq 1$ and this ensures that the series converges in the disc.

To see that $f \in H_1^\infty$ we make a long computation:

$$1 - \overline{f(z)}f(z) = 1 - (a + zB(I - zD)^{-1}C)^*(a + zB(I - zD)^{-1}C)$$

$$= \ldots$$

$$= C^* \left((I - zD)^{-1}\right)^* \left[1 - |z|^2\right] (I - zD)^{-1}C \geq 0.$$

(To get from the first line to the second one we used the fact that

$$V^*V = \begin{pmatrix} 1 & 0 \\ 0 & I \end{pmatrix},$$

which gives $\overline{a}a + C^*C = 1$, $\overline{a}B + C^*D = 0$ and $B^*B + D^*D = I$.) This implies that $|f(z)|^2 \leq 1$ for all $z \in \mathbb{D}$, in other words $\|f\|_\infty \leq 1$, and the proof is complete. $\qquad\square$

**Proof of Pick's theorem**

We now complete the proof of Pick's interpolation theorem. Corollary 6.3.16 takes care of one direction: positivity of the Pick matrix is a necessary condition for the existence of an interpolating $f \in H^\infty(\mathbb{D})$ with norm $\|f\|_\infty \leq 1$. It remains to show that the positivity condition is sufficient; this will be achieved in Theorem 6.3.19 below, which gives more precise information. We shall require the following lemma.

**Lemma 6.3.18.** *Let $k : X \times X \to \mathbb{C}$ be a positive semidefinite kernel. Then there exists a Hilbert space $K$ and a function $F : X \to K$ such that $k(x, y) = \langle F(x), F(y) \rangle$ for all $x, y \in X$. If $X$ is a set with $n$ points, then the dimension of $K$ can be chosen to be the rank of $k$ when considered as a positive semidefinite $n \times n$ matrix.*

*Proof.* Define a kernel $\tilde{k}(x, y) = k(y, x)$. If $k$ is positive semidefinite, then so is $\tilde{k}$. Let $K$ be the reproducing kernel Hilbert space associated with $\tilde{k}$ as in Theorem 6.1.6. Define $F : X \to K$ by $F(x) = \tilde{k}_x$. Then we have

$$\langle F(x), F(y) \rangle = \tilde{k}(y, x) = k(x, y).$$

$\square$

Recall that a **rational function** is a function of the form $f = \frac{p}{q}$ where $p$ and $q$ are polynomials. By the **degree** of a rational function $f = \frac{p}{q}$ (assumed to be in reduced form) we mean $\max\{\deg(p), \deg(q)\}$.

**Theorem 6.3.19.** *Let $z_1, \ldots, z_n \in \mathbb{D}$ and $w_1, \ldots, w_n \in \mathbb{C}$ be given. If the Pick matrix*

$$\left( \frac{1 - w_i \overline{w_j}}{1 - z_i \overline{z_j}} \right)_{i,j=1}^n$$

*is positive semidefinite and has rank $m$, then there exists a rational function $f$ of degree at most $m$ such that $\|f\|_\infty \leq 1$ and $f(z_i) = w_i$, for all $i = 1, \ldots, n$.*

A consequence of the boundedness of $f$ is that the poles of $f$ lie away from the closed unit disc.

*Proof.* By Lemma 6.3.18, there are $F_1, \ldots, F_n \in \mathbb{C}^m$ such that $\frac{1 - w_i \overline{w_j}}{1 - z_i \overline{z_j}} = \langle F_i, F_j \rangle$ for all $i, j$. We can rewrite this identity in the following form

$$1 + \langle z_i F_i, z_j F_j \rangle = w_i \overline{w_j} + \langle F_i, F_j \rangle.$$

By Exercise 5.7.4, it is possible to define an isometry $V$ from

$$\text{span}\{(1, z_i F_i) : i = 1, \ldots, n\} \subseteq \mathbb{C} \oplus \mathbb{C}^m$$

into

$$\text{span}\{(w_i, F_i) : i = 1, \ldots, n\} \subseteq \mathbb{C} \oplus \mathbb{C}^m,$$

by sending $(1, z_i F_i)$ to $(w_i, F_i)$ and extending linearly. Since isometric subspaces have equal dimension, we can extend $V$ to an isometry $V : \mathbb{C} \oplus \mathbb{C}^m \to \mathbb{C} \oplus \mathbb{C}^m$.

The realization formula suggests a way to write down a function $f \in H_1^\infty$ given an isometric matrix in block form. So let us write

$$V = \begin{pmatrix} a & B \\ C & D \end{pmatrix},$$

where the decomposition is according to $\mathbb{C} \oplus \mathbb{C}^m$, and define $f(z) = a + zB(I - zD)^{-1}C$. By Theorem 6.3.17, $f \in H_1^\infty$, and it is rational because the inverse of a matrix is given by rational formulas involving the entries. The degree of $f$ is evidently not greater than $m$. It remains to show that $f$ interpolates the data.

Fix $i \in \{1, \ldots, n\}$. From the definition of $V$ we have

$$\begin{pmatrix} a & B \\ C & D \end{pmatrix} \begin{pmatrix} 1 \\ z_i F_i \end{pmatrix} = \begin{pmatrix} w_i \\ F_i \end{pmatrix}.$$

The first row gives $a + z_i B F_i = w_i$. The second row gives $C + z_i D F_i = F_i$, so solving for $F_i$ we obtain $F_i = (I - z_i D)^{-1} C$ (where the inverse is legal, because $\|z_i D\| < 1$). Plugging this in the first row gives

$$w_i = a + z_i B F_i = a + z_i B (I - z_i D)^{-1} C = f(z_i),$$

thus $f$ interpolates the data, and the proof is complete.  $\square$

**Remark 6.3.20.** The argument that we used to prove Pick's theorem has a cute name: the *lurking isometry* argument. It is useful in many contexts.

**Exercise 6.3.21.** Complete the proof of Theorem 6.3.17. (**Hint:** use a lurking isometry argument.)

---

## 6.4  Additional exercises

**Exercise 6.4.1.** Rewrite the proof of Theorem 6.1.6 so that it involves no quotient-taking and no abstract completions.

**Exercise 6.4.2.** Prove that

$$\left\{ \frac{1}{\sqrt{\pi}}, \sqrt{\frac{2}{\pi}} z, \sqrt{\frac{3}{\pi}} z^2, \ldots, \sqrt{\frac{n+1}{\pi}} z^n, \ldots \right\}$$

is an orthonormal basis for $L_a^2(\mathbb{D})$.

**Exercise 6.4.3.** What is the multiplier algebra of the Bergman space $L_a^2(\mathbb{D})$?

**Exercise 6.4.4.** Let $H$ be a Hilbert function space on a set $X$. A *composition operator* on $H$ is a map of the form $h \mapsto h \circ \sigma$, where $\sigma : X \mapsto X$ is some map.

1. Suppose that $T : h \mapsto h \circ \sigma$ is a bounded composition operator. Find a formula for $T^*$.

2. True or false: there exists a composition operator whose adjoint is not a composition operator.

3. Let $\sigma$ be an analytic map defined on an open neighborhood of the unit disc, and assume that $\sigma$ maps $\mathbb{D}$ into $\mathbb{D}$. Show that $\sigma$ induces a bounded composition operator on $L_a^2(\mathbb{D})$. Does it also induce a bounded composition operator on $H^2(\mathbb{D})$?

**Exercise 6.4.5.** Two Hilbert function spaces $H$ on $X$ (with kernel $k^H$) and $G$ on $W$ (with kernel $k^G$) are said to be *isomorphic as Hilbert function spaces* if there is a bijection $\sigma : X \to W$ such that

$$k^H(x, y) = k^G(\sigma(x), \sigma(y)) \quad \text{for all } x, y \in X.$$

In other words, $H$ and $G$ are isomorphic as Hilbert function spaces if the map

$$k_x^H \mapsto k_{\sigma(x)}^G$$

extends to a unitary map between $H$ and $G$.

1. Prove that if $H$ and $G$ are isomorphic as Hilbert function spaces, then their multiplier algebras are isometrically isomorphic (i.e., there exists an algebraic isomorphism between the algebras that is also norm preserving).

2. Are the Bergman space and the Hardy space isomorphic as Hilbert function spaces?

**Exercise 6.4.6.** For a Hilbert space $H$, and a set $\mathcal{S} \subseteq B(H)$, we define

$$\mathcal{S}' = \{T \in B(H) : ST = TS \text{ for all } S \in \mathcal{S}\}.$$

In this exercise, we will consider the multiplier algebra of a Hilbert function space as a subalgebra of $B(H)$.

1. Prove that $\text{Mult}(H^2(\mathbb{D}))' = \text{Mult}(H^2(\mathbb{D}))$.

2. Is it true that $\text{Mult}(L_a^2(\mathbb{D}))' = \text{Mult}(L_a^2(\mathbb{D}))$?

3. Is it true that $\text{Mult}(H)' = \text{Mult}(H)$ for all Hilbert function spaces?

**Exercise 6.4.7** (Agler[2]). Let $G = C^1([0,1])$ (the continuously differentiable functions on the interval $[0,1]$) with inner product

$$\langle f, g \rangle = \int_0^1 f(t)\overline{g}(t)dt + \int_0^1 f'(t)\overline{g'}(t)dt.$$

1. Prove that point evaluation at any $x \in [0,1]$ is a bounded functional on $G$.

2. Let $H$ be the completion of $G$. Explain why $H$ can be considered as a concrete space of functions on $[0,1]$, in which point evaluation is bounded. ($H$ is actually equal to the *Sobolev space* $H^1[0,1]$ of absolutely continuous functions on $[0,1]$ with derivatives in $L^2[0,1]$.)

3. Find the kernel of $H$.

4. Prove that every $g \in G$ gives rise to a multiplier on $H$. Show that $\|M_g\| > \sup_{t \in [0,1]} |g(t)|$, unless $g$ is constant.

5. (Harder) Show that $\mathrm{Mult}(H) = H$ (as a set of functions).

---

[2]Agler also proves that a version of Pick's theorem holds true in this space, see [1].

# Chapter 7

## Banach spaces

Up to this point, we discussed mainly Hilbert spaces. We defined the space $B(H,K)$ of bounded operators between two Hilbert spaces. On $B(H,K)$ we defined a norm,

$$\|T\| = \sup_{\|h\|=1} \|Th\|. \tag{7.1}$$

This makes $B(H,K)$ a normed space, and we will soon see that it is complete with respect to this norm. Thus, even if we are only interested about questions in Hilbert spaces (and we are certainly not!), we are led nonetheless to consider other kinds of normed spaces. This is a convenient point to switch to a somewhat more general point of view, and enlarge our scope to include all complete normed spaces — Banach spaces.

## 7.1 Banach spaces

### 7.1.1 Basic definitions and examples

We will continue to consider linear spaces over $\mathbb{R}$ or $\mathbb{C}$, and we keep our convention that whenever we need to refer to the scalar field we will take the complex numbers.

**Definition 7.1.1.** A *norm* on a linear space $X$ is a function $\|\cdot\| : X \to [0,\infty)$ such that

1. $\|x\| \geq 0$ for all $x \in X$, and $\|x\| = 0$ implies that $x = 0$.

2. $\|ax\| = |a|\|x\|$ for all $x \in X$ and all $a \in \mathbb{C}$.

3. $\|x+y\| \leq \|x\| + \|y\|$ for all $x,y \in X$.

A *normed space* is simply a linear space that has a norm defined on it.

**Definition 7.1.2.** If $X$ is a normed space, then the *closed unit ball* of $X$, denoted by $X_1$, is the set

$$X_1 = \{x \in X : \|x\| \leq 1\}.$$

Every normed space is a metric space, where the metric is induced by the norm $d(x, y) = \|x - y\|$. A normed space is said to have a certain property if the underlying metric space has this property. For example, a normed space is said to be **separable** if it is separable as a metric space, and it is said to be **complete** if it is complete as a metric space.

**Definition 7.1.3.** A **Banach space** is a normed space which is complete.

**Example 7.1.4.** Every Hilbert space is a Banach space.

**Example 7.1.5.** Let $X$ be a topological space. The space $C_b(X)$ of bounded continuous functions equipped with the sup norm $\|f\|_\infty = \sup_{x \in X} |f(x)|$ is a Banach space. In particular, if $X$ is a compact space, then $C(X)$ equipped with the norm $\| \cdot \|_\infty$ is a Banach space.

Let $p \in [1, \infty)$. For any finite or infinite sequence $x = (x_k)_k$ we define

$$\|x\|_p = \left( \sum_k |x_k|^p \right)^{1/p}.$$

We also define

$$\|x\|_\infty = \sup_k |x_k|.$$

An important family of finite dimensional Banach spaces is given by the spaces commonly denoted by $\ell_n^p$ (for $p \in [1, \infty]$ and $n \in \mathbb{N}$), which are the vector space $\mathbb{C}^n$ endowed with the norm $\| \cdot \|_p$.

Not less important are the infinite dimensional $\ell^p$ spaces $(p \in [1, \infty])$, given by

$$\ell^p = \{x = (x_k)_{k=0}^\infty \in \mathbb{C}^{\mathbb{N}} : \|x\|_p < \infty\}.$$

For $p = 2$ we get the Hilbert space $\ell^2$ which we have already encountered. For $p = 1$ or $p = \infty$, it is clear that $\| \cdot \|_p$ is a norm, but for $p \notin \{1, 2, \infty\}$ it is not apparent that $\| \cdot \|_p$ is a norm. Our present goal is to prove that it is.

**Definition 7.1.6.** Two extended real numbers $p, q \in [1, \infty]$ are said to be **conjugate exponents** if

$$\frac{1}{p} + \frac{1}{q} = 1.$$

If $p = 1$, then we understand this to mean that $q = \infty$, and vice versa.

**Lemma 7.1.7** (Hölder's inequality). *Let $p, q \in [1, \infty]$ be conjugate exponents. Then for any two (finite or infinite) sequences $x_1, x_2, \ldots$ and $y_1, y_2, \ldots$*

$$\sum_k |x_k y_k| \leq \|x\|_p \|y\|_q.$$

*Proof.* The heart of the matter is to prove the inequality for finite sequences. Pushing the result to infinite sequences does not require any clever idea, and is left to the reader (no offense). Therefore, we need to prove that for every $x = (x_k)_{k=1}^n$ and $y = (y_k)_{k=1}^n$ in $\mathbb{C}^n$,

$$\sum_{k=1}^n |x_k y_k| \leq \left(\sum_{k=1}^n |x_k|^p\right)^{1/p} \left(\sum_{k=1}^n |y_k|^q\right)^{1/q}. \qquad (7.2)$$

The case $p = 1$ (or $p = \infty$) is immediate. The right-hand side of (7.2) is continuous in $p$ when $x$ and $y$ are held fixed, so it is enough to verify the inequality for a dense set of values of $p$ in $(1, \infty)$. Define

$$S = \left\{\frac{1}{p} \in (0,1) \,\middle|\, p \text{ satisfies Equation (7.2) for all } x, y \in \mathbb{C}^n\right\}.$$

Now our task reduces to that of showing that $S$ is dense in $(0,1)$. By the Cauchy-Schwarz inequality, we know that $\frac{1}{2} \in S$. Also, the roles of $p$ and $q$ are interchangeable, so $\frac{1}{p} \in S$ if and only if $1 - \frac{1}{p} \in S$.

Set $a = \frac{q}{2p+q}$ ($a$ is chosen to be the solution to $2ap = (1-a)q$, we will use this soon). Now, if $\frac{1}{p} \in S$, we apply (7.2) to the sequences $(|x_k||y_k|^a)_k$ and $(|y_k|^{1-a})_k$, and then we use the Cauchy-Schwarz inequality, to obtain

$$\sum_{k=1}^n |x_k y_k| = \sum_{k=1}^n |x_k||y_k|^a |y_k|^{1-a}$$

$$\leq \left(\sum_{k=1}^n |x_k|^p |y_k|^{ap}\right)^{1/p} \left(\sum_{k=1}^n |y_k|^{(1-a)q}\right)^{1/q}$$

$$\leq \left(\left(\sum_{k=1}^n |x_k|^{2p}\right)^{1/2} \left(\sum_{k=1}^n |y_k|^{2ap}\right)^{1/2}\right)^{1/p} \left(\sum_{k=1}^n |y_k|^{(1-a)q}\right)^{1/q}$$

$$= \left(\sum_{k=1}^n |x_k|^{p'}\right)^{\frac{1}{p'}} \left(\sum_{k=1}^n |y_k|^{q'}\right)^{\frac{1}{q'}},$$

where $\frac{1}{p'} = \frac{1}{2p}$ and $\frac{1}{q'} = \frac{1}{2p} + \frac{1}{q}$. Therefore, if $s = \frac{1}{p} \in S$, then $\frac{s}{2} = \frac{1}{2p} \in S$; and if $s = \frac{1}{q} \in S$, then $\frac{s+1}{2} = \frac{1}{2}\frac{1}{q} + \frac{1}{2} = \frac{1}{q} + \frac{1}{2}\frac{1}{p}$ is also in $S$.

Since $\frac{1}{2}$ is known to be in $S$, it follows by induction that for every $n \in \mathbb{N}$ and $m \in \{1, 2, \ldots, 2^n - 1\}$, the fraction $\frac{m}{2^n}$ is in $S$. Hence $S$ is dense in $(0,1)$, and the proof is complete. □

**Remark 7.1.8.** It is sometimes convenient to invoke Hölder's inequality in the form

$$\left|\sum_k x_k y_k\right| \leq \left(\sum_k |x_k|^p\right)^{1/p} \left(\sum_k |y_k|^q\right)^{1/q},$$

which obviously follows from the one we proved. In fact, this form is equivalent (prove it).

**Lemma 7.1.9** (Minkowski's inequality). *For every $p \in [1, \infty]$ and every two (finite or infinite) sequences $x_1, x_2, \ldots$ and $y_1, y_2, \ldots$*

$$\|x + y\|_p \leq \|x\|_p + \|y\|_p.$$

*In particular, if $x, y \in \ell^p$, then $x + y \in \ell^p$.*

*Proof.* Again, the cases $p = 1, \infty$ are easy, and the result for infinite sequences is a consequence of the result for finite sequences. Therefore, let $p \in (1, \infty)$, and let $x, y \in \mathbb{C}^n$. With $q = \frac{p}{p-1}$, we use Hölder's inequality and the identity $q(p-1) = p$ to find

$$\sum_k |x_k + y_k|^p = \sum_k |x_k + y_k||x_k + y_k|^{p-1}$$

$$\leq \sum_k |x_k||x_k + y_k|^{p-1} + \sum_k |y_k||x_k + y_k|^{p-1}$$

$$\leq \|x\|_p \left( \sum_k |x_k + y_k|^p \right)^{1/q} + \|y\|_p \left( \sum_k |x_k + y_k|^p \right)^{1/q}$$

$$= (\|x\|_p + \|y\|_p) \|x + y\|_p^{p-1}.$$

Minkowski's inequality follows. $\square$

**Theorem 7.1.10.** *For every $p \in [1, \infty]$ and all $n \in \mathbb{N}$, the spaces $\ell_n^p$ and $\ell^p$ are Banach spaces.*

*Proof.* Let us focus on the case of infinite sequences. By Minkowski's inequality, $\ell^p$ is a vector space and $\|\cdot\|_p$ is a norm. For $p \neq \infty$, completeness is established in precisely the same way that is was for the space $\ell^2$, by replacing 2 by $p$. The case $p = \infty$ follows from Example 7.1.5. $\square$

**Exercise 7.1.11.** Prove directly that $\ell^\infty$ is a Banach space.

The category of normed spaces is much richer than that of Hilbert spaces. Unless $p = 2$, there does not exist an inner product on $\ell^p$ (or on $\ell_n^p$) that gives rise to the $\|\cdot\|_p$ norm. Norms that are induced by an inner product are exceptional.

**Exercise 7.1.12** (Hard[1]). A norm on a normed space $X$ is induced by an inner product if and only if it satisfies the parallelogram law:

$$\|x + y\|^2 + \|x - y\|^2 = 2\|x\|^2 + 2\|y\|^2, \quad \text{for all } x, y \in X.$$

Using the result in the exercise, it is not hard to show that some very frequently occurring norms, such as the $\|\cdot\|_p$ norms or the operator norm on $B(H)$, are not induced by inner products (note that for showing that a norm is not induced by an inner product, only the easy direction is required).

[1]See [15].

## 7.1.2 Completion

**Theorem 7.1.13.** *Every normed space $X$ can be embedded isometrically in a unique Banach space which contains $X$ as a dense subset.*

The proof of Theorem 7.1.13 is just like that of Theorem 2.3.1 (the completion Theorem in Hilbert spaces), only simpler. The unique Banach space containing $X$ is called the **completion** of $X$.

**Exercise 7.1.14.** Let $\ell_0$ denote the vector space of all finitely supported sequences. For $1 \leq p \leq \infty$, equip $\ell_0$ with the $\|\cdot\|_p$ norm. Compute the completion of $\ell_0$ with respect to $\|\cdot\|_p$.

**Exercise 7.1.15.** A function $f \in C(\mathbb{R}^k)$ is said to have **compact support** if the support of $f$, defined to be $\mathrm{supp}(f) = \overline{\{x : f(x) \neq 0\}}$, is a compact subset of $\mathbb{R}^k$. A function $f$ is said to **vanish at infinity** if for all $\epsilon > 0$ there is a compact $K$ such that $|f(x)| < \epsilon$ for all $x \notin K$. Let $C_c(\mathbb{R}^k)$ denote the space of all compactly supported continuous functions, and let $C_0(\mathbb{R}^k)$ denote the space of all continuous functions vanishing at infinity. Prove that $C_c(\mathbb{R}^k)$ is an incomplete normed space, and that the completion of $C_c(\mathbb{R}^k)$ is $C_0(\mathbb{R}^k)$.

**Example 7.1.16.** Let $D$ be an nice subset in $\mathbb{R}^k$ (for example, an open set, or a square $[0,1]^k$, etc.). Let $C_c(D)$ denote the continuous functions with compact support in $D$. Define a norm on $C_c(D)$ by $\|f\|_p = (\int_D |f(x)|^p dx)^{1/p}$. The fact that this is a norm depends on the Minkowski inequality for integrals (see Exercise 7.5.6). Then $L^p(D) = L^p(D, dx)$ can be identified with (and can also be defined as) the completion of $C_c(D)$ with respect to this norm.

Note that if $p = 2$ and $D = [a, b]$, then thanks to Exercise 2.4.6 we get the same definition of $L^2[a, b]$ as we introduced in Section 2.4.

**Exercise 7.1.17.** Prove that $L^p(D)$ is separable (i.e., that $L^p(D)$ contains a countable dense subset) for all $p \in [1, \infty)$. Prove that $\ell^p$ is separable for $p \in [1, \infty)$ and nonseparable for $p = \infty$.

**Example 7.1.18.** The previous example fits into a larger family of spaces, which are defined in terms of measure theory. If $(X, \mu)$ is a measure space and $p \geq 1$, then one may define the Lebesgue spaces $L^p(X, \mu)$ as the space of all measurable functions on $X$ for which $\int_X |f|^p d\mu < \infty$. If $X = \{1, 2, \dots, \}$ and $\mu$ is the counting measure, then we get the space $\ell^p$ (the counting measure is the measure that assigns to each set $S$ the value $\mu(S) = $ the number of elements in $S$). The reader should consult [11] and [29] to learn about $L^p$ spaces.

## 7.2   Bounded operators

### 7.2.1   Definition and completeness

Let $X$ and $Y$ be normed spaces. We define the space of bounded linear operators from $X$ to $Y$ just as we have in the Hilbert space case.

**Definition 7.2.1.** Let $X$ and $Y$ be normed spaces. A linear operator $T :$ $X \to Y$ is said to be **bounded** if the **operator norm** $\|T\|$ of $T$, defined as

$$\|T\| = \sup_{\|x\|=1} \|Tx\|,$$

satisfies $\|T\| < \infty$. The space of all bounded operators from $X$ to $Y$ is denoted by $B(X, Y)$. If $Y = X$ we abbreviate $B(X) = B(X, X)$.

It follows immediately that $\|Tx\| \leq \|T\| \|x\|$ for all $x \in X$. When equipped with the operator norm $\|T\|$, the space $B(X, Y)$ becomes a normed space.

**Exercise 7.2.2.** Prove that $\|T\| = \sup_{\|x\|=1} \|Tx\|$ is a norm on $B(X, Y)$.

**Exercise 7.2.3.** Show that for every $T \in B(X, Y)$,

$$\|T\| = \sup_{\|x\| \leq 1} \|Tx\| = \sup_{\|x\| < 1} \|Tx\|.$$

We have the notions of **kernel** and **image** of operators between Banach spaces, defined identically as in the case of inner product spaces (recall Definition 5.1.8). The proof of the following proposition is identical to proof of Proposition 5.1.4.

**Proposition 7.2.4.** *For a linear transformation* $T : X \to Y$ *mapping between two normed spaces, the following are equivalent:*

1. *T is bounded.*

2. *T is continuous.*

3. *T is continuous at some* $x_0 \in X$.

**Proposition 7.2.5.** *If* $Y$ *is a Banach space, then* $B(X, Y)$ *is also a Banach space.*

*Proof.* Suppose that $\{A_n\}$ is a Cauchy sequence in $B(X, Y)$. For all $x \in X$, $\{A_n x\}$ is a Cauchy sequence in the complete space $Y$. Define $A(x) = \lim A_n x$. It is easy to see that the map $A$ is linear and bounded. Even though we wrote down $Ax = \lim A_n x$, we still did not show that $A_n \to A$ in $B(X, Y)$, because this involves showing that $\|A_n - A\| \to 0$. If $n_0$ is chosen large enough so that $\|A_n - A_m\|$ is smaller than $\epsilon$ for $m, n \geq n_0$, then for all $x \in X$,

$$\|(A - A_m)x\| = \lim_{n \to \infty} \|(A_n - A_m)x\| \leq \epsilon \|x\|.$$

Thus $\|A - A_m\| \leq \epsilon$ for $m \geq n_0$, hence $A_n \to A$ in norm.    $\square$

## 7.2.2 Examples of bounded operators

We let $I_X$ or $I$ or 1 denote the identity mapping from $X$ to itself. Clearly $I \in B(X)$, and so is every scalar multiple of it.

**Definition 7.2.6.** An operator $T \in B(X,Y)$ is said to be a **contraction** if $\|T\| \leq 1$. It is said to be **isometric** (or an **isometry**) if $\|Tx\| = \|x\|$ for all $x \in X$.

**Example 7.2.7.** Let $X = C(K)$, where $K$ is a compact topological space. For every $g \in X$, we can define the multiplication operator

$$M_g : f \mapsto gf.$$

Then $M_g \in B(X)$, and $\|M_g\| = \|g\|_\infty$.

**Example 7.2.8.** Let $K$ be a compact topological space and let $\alpha : K \to K$ be a continuous function. Define a transformation $U : C(K) \to C(K)$ by $Uf = f \circ \alpha$. Then $U$ is a contraction operator on $C(K)$.

The above examples were given by a concrete formula defined on the entire space $X$. In practice many operators are defined first on a dense linear subspace, and extended to the entire space $X$ by continuity.

**Exercise 7.2.9.** Let $D$ be a dense linear subspace of a normed space $X$. Let $Y$ be a Banach space, and let $T \in B(D,Y)$. Show that there is a unique $\tilde{T} \in B(X,Y)$ such that $\tilde{T}\big|_D = T$ and $\|\tilde{T}\| = \|T\|$. Moreover, prove that if $T$ is isometric, then $\tilde{T}$ is, too.

**Example 7.2.10.** Let $\alpha : [0,1] \to [0,1]$ be a piecewise continuous function. Assume further that $\alpha$ is **measure preserving**. By this we mean that for all $f \in PC[0,1]$,

$$\int_0^1 f \circ \alpha(x)dx = \int_0^1 f(x)dx. \tag{7.3}$$

(We will see in Section 8.2 nontrivial examples of measure preserving maps.) Define a transformation $U : C([0,1]) \to PC[0,1]$ by $Uf = f \circ \alpha$. By our assumptions on $\alpha$, $U$ is well-defined, linear and isometric on $C([0,1])$, which is a dense subspace of $L^p[0,1]$ for all $1 \leq p < \infty$. By the above exercise, $U$ extends in a unique way to an isometry $U : L^p[0,1] \to L^p[0,1]$. We may continue to write $Uf = f \circ \alpha$ even for $f \in L^p[0,1]$, giving thereby a meaning to the notion of the phrase "composing a function $f \in L^p$ with $\alpha$". Composition operators such as $U$ are of much interest, and we will return to this subject in Section 8.2.

## 7.3   The dual space

### 7.3.1   Definition and examples

**Definition 7.3.1.** Let $X$ be a normed space. A linear map from $X$ into the scalar field is called a **linear functional.** We let $X^*$ denote the space of all bounded linear functionals on $X$. The space $X^*$ is called the **dual space** (or the **conjugate space**) of $X$.

In some parts of the literature, the dual space is denoted by $X'$.

**Remark 7.3.2.** The dual space $X^*$, being a space of bounded operators on $X$, is a normed space with the norm

$$\|f\| = \sup_{x \in X_1} |f(x)| \quad , \quad f \in X^*.$$

By Proposition 7.2.5, $X^*$ is complete.

**Example 7.3.3.** Every Hilbert space can be identified (isometrically and anti-linearly) with its dual in the following way

$$H^* = \{\langle \cdot, g \rangle : g \in H\}.$$

**Example 7.3.4.** The dual of $\ell^1$ can be isometrically identified with $\ell^\infty$; this is usually summarized by writing $(\ell^1)^* = \ell^\infty$.

To be precise, every $b \in \ell^\infty$ gives rise to a functional $\Phi_b \in (\ell^1)^*$ by way of $\Phi_b(a) = \sum_n a_n b_n$. Clearly $|\Phi_b(a)| \leq \|b\|_\infty \|a\|_1$ for every $a \in \ell^1$, so $\Phi_b$ is well-defined, bounded, and $\|\Phi_b\| \leq \|b\|_\infty$. To complete the identification of $(\ell^1)^*$ with $\ell^\infty$, we will show that the map $b \mapsto \Phi_b$ is also isometric and surjective.

For every $n$, let $e_n$ denote the element of $\ell^1$ with 1 in the $n$th slot and 0s elsewhere. Then it is easy to see that for every $a = (a_n)_n \in \ell^1$, we have the convergent sum

$$a = \lim_{N \to \infty} \sum_{n=0}^{N} a_n e_n = \sum_{n \in \mathbb{N}} a_n e_n.$$

Given $\Psi \in (\ell^1)^*$, we define a sequence $b = (b_n)_n$ by $b_n = \Psi(e_n)$. Then $\|b\|_\infty \leq \|\Psi\|$. By linearity and continuity,

$$\Psi(a) = \lim_{N \to \infty} \sum_{n=0}^{N} a_n \Psi(e_n) = \lim_{N \to \infty} \sum_{n=0}^{N} a_n b_n = \Phi_b(a),$$

whence, $\Psi = \Phi_b$. Thus $b \mapsto \Phi_b$ is surjective. Finally, $\|b\|_\infty \leq \|\Psi\| = \|\Phi_b\| \leq \|b\|_\infty$, so $b \mapsto \Phi_b$ is isometric.

**Example 7.3.5.** Similarly to the above example, one can show that every $b \in \ell^1$ gives rise to a bounded functional $\Gamma_b$ on $\ell^\infty$ given by

$$\Gamma_b(a) = \sum a_n b_n.$$

The map $b \mapsto \Gamma_b$ is an isometric linear map of $\ell^1$ into $(\ell^\infty)^*$. We currently do not have enough tools to determine whether this map is surjective (this issue will be taken up again in Chapter 13; a reader who will solve this problem before we get to that chapter deserves bonus points).

**Exercise 7.3.6.** Show that if $1 < p, q < \infty$ are conjugate exponents, then $(\ell^p)^* = \ell^q$, where the isomorphism is similar to the one given in Example 7.3.4.

### 7.3.2 Weak convergence

In this section we introduce an important notion in Banach space theory, that of weak convergence. Although the notion makes sense in any Banach space, we will only prove significant results regarding this notion in the Hilbert space setting.

**Definition 7.3.7.** A sequence $\{x_n\}_{n=1}^\infty$ in a normed space $X$ is said to be *weakly convergent* to $x \in X$ if for every $f \in X^*$,

$$\lim_n f(x_n) = f(x).$$

If $\{x_n\}_{n=1}^\infty$ weakly converges to $x$, then we write $x_n \rightharpoonup x$.

The usual (with respect to norm) convergence in a normed space is sometimes called *strong convergence* in discussions where confusion may be caused. It is clear from the definition that strong convergence implies weak convergence, but the converse does not hold.

**Example 7.3.8.** Let $H$ be a Hilbert space and let $\{e_n\}_{n=1}^\infty$ be an orthonormal sequence. By Bessel's inequality, $\langle e_n, h \rangle \to 0$ for every $h \in H$, thus $e_n \rightharpoonup 0$. Clearly, the sequence does not converge strongly to 0. Incidently, note that the norm is "not continuous with respect to weak convergence".

**Exercise 7.3.9.** Prove that the limit of a weakly convergent sequence in a Hilbert space is unique. What can you say about the uniqueness of the weak limit in a Banach space?

One may go on to define the so-called *weak topology* on a normed space $X$, but care is needed, since this topology is not determined in general by convergent sequences. As the above example hints, the weak and norm topologies differ in general, but they are very closely related — for example convex sets are closed in one topology if and only if they are closed in the other. In the setting of Hilbert spaces, this fact follows from the following, stronger result.

**Theorem 7.3.10** (Banach-Saks theorem). *Let $H$ be a Hilbert space, and let $\{h_n\}_{n=1}^{\infty}$ be a sequence such that $h_n \rightharpoonup h$. Then there exists a subsequence $\{h_{n_k}\}_{k=1}^{\infty}$ whose Cesàro means converge to $h$ in norm, i.e.,*

$$\frac{1}{N} \sum_{k=1}^{N} h_{n_k} \to h.$$

We will prove this theorem below. But first, we need another nontrivial result (see Exercise 7.5.17 for a generalization).

**Theorem 7.3.11** (Principle of uniform boundedness (baby version)). *Let $\{h_n\}$ be a weakly convergent sequence in a Hilbert space. Then $\{h_n\}$ is bounded: $\sup_n \|h_n\| < \infty$.*

*Proof.* Suppose that $\{h_n\}$ is a weakly convergent sequence. For every $N \in \mathbb{N}$, we define

$$F_N = \{g \in H : |\langle g, h_n \rangle| \leq N\|g\| \text{ for all } n\}.$$

Every $F_N$ is closed (it is an intersection over $n$ of closed sets). As $\{\langle g, h_n \rangle\}_n$ is a convergent sequence for every fixed $g$, it is bounded, thus $H = \cup_N F_N$. The space $H$ is complete, therefore Baire's category theorem (Theorem A.2.17) implies that there is some $N$ such that $F_N$ has nonempty interior. Suppose that $B_r(g) \subset F_N$. This means that

$$\|g - f\| < r \text{ implies that } |\langle f, h_n \rangle| \leq N\|f\| \text{ for all } n.$$

We thus have that if $\|f\| < r$, then for all $n$,

$$|\langle f, h_n \rangle| \leq |\langle g + f, h_n \rangle| + |\langle g, h_n \rangle| \leq N(\|f + g\| + \|g\|).$$

Taking the supremum over all $f$ with $\|f\| < r$, it follows that $\|h_n\| \leq r^{-1}(2\|g\| + r)$ for all $n$. $\qquad\square$

**Proof of the Banach-Saks Theorem.** We may assume that $h_n \rightharpoonup 0$ (otherwise, replace $h_n$ with $h_n - h$). We need to find a subsequence $\{h_{n_k}\}$ such that

$$\frac{1}{N} \sum_{k=1}^{N} h_{n_k} \to 0.$$

Choose a subsequence such that for every $k$, we have $|\langle h_{n_i}, h_{n_k} \rangle| < 2^{-k}$ for all $i = 1, 2, \ldots, k - 1$. But then

$$\left\| \frac{1}{N} \sum_{k=1}^{N} h_{n_k} \right\|^2 = \frac{1}{N^2} \sum_{k=1}^{N} \|h_{n_k}\|^2 + \frac{1}{N^2} \sum_{1 \leq k \neq j \leq N} \langle h_{n_j}, h_{n_k} \rangle$$

$$\leq \frac{\sup_n \|h_n\|^2}{N} + \frac{2}{N^2} \sum_{k=2}^{N} \sum_{j=1}^{k-1} 2^{-k}.$$

Thanks to the principle of uniform boundedness, the right-hand side converges to 0. $\qquad\square$

The following exercises explore additional important facts about weak convergence.

**Exercise 7.3.12.** Prove that every bounded sequence in a Hilbert space has a weakly convergent subsequence. This property of Hilbert spaces is sometimes referred to as **weak sequential compactness** of the closed unit ball.

**Exercise 7.3.13.** Let $A$ be linear operator between Hilbert spaces, and assume that $A$ is **weakly sequentially continuous**, in the sense that

$$h_n \rightharpoonup h \quad \text{implies that} \quad Ah_n \rightharpoonup Ah.$$

Prove that $A$ is bounded. Conversely, prove that a bounded operator is weakly sequentially continuous.

**Exercise 7.3.14.** Suppose that $\{h_n\}$ is a sequence in a Hilbert space $H$ such that for every $g \in H$, the sequence $\{\langle h_n, g \rangle\}$ is convergent. Then $\{h_n\}$ is a weakly convergent sequence, that is, there is some $h \in H$ such that

$$\langle h_n, g \rangle \to \langle h, g \rangle$$

for all $g \in H$. This property is called **weak sequential completeness**. (**Hint:** First modify the proof of Theorem 7.3.11 to show that $\{h_n\}$ is bounded.)

---

## 7.4 *Topological vector spaces

Before moving on, it should be mentioned that Banach spaces, though by far more general than Hilbert spaces, do not cover all the kinds of topological vector spaces that arise in analysis or in applications.

Consider for example the space $C(\Omega)$, where $\Omega \subseteq \mathbb{R}^k$. When $\Omega$ is not compact there exist unbounded continuous functions on it. Thus one cannot define the supremum norm for all elements in $C(\Omega)$. However, for every compact subset $K$ of $\Omega$ and every $f \in C(\Omega)$, we can define $\|f\|_K = \sup_{x \in K} |f(x)|$. If $K_1, K_2, \ldots$ is a sequence of compact sets such that $\Omega = \cup_n K_n$ and $K_n \subset \operatorname{int} K_{n+1}$ for all $n$, then we may define a measure of distance between elements in $C(\Omega)$ by

$$d(f, g) = \sum_n 2^{-n} \frac{\|f - g\|_{K_n}}{1 + \|f - g\|_{K_n}}.$$

One shows that $d$ is a complete metric on $C(\Omega)$, and that a sequence $\{f_n\}$

converges to $f$ in $(C(\Omega), d)$ if and only if $f_n \to f$ uniformly on every compact subset of $\Omega$.

It can be shown that there is no norm one can define on $C(\Omega)$ that induces this metric. Other examples of interesting function spaces which are not linearly homeomorphic to Banach spaces are $C^\infty(\Omega)$ (the space of infinitely differentiable functions on $\Omega$) or $\mathcal{O}(\Omega)$ (the space of analytic functions on $\Omega$) when $\Omega$ is a domain in $\mathbb{C}$.

A vector space which is also a complete metric space with a translation invariant metric which has a convex local base at 0 is called a **Fréchet space**. There are even more general topological vector spaces of interest in analysis, especially **locally convex topological vector spaces**. These spaces are beyond the scope of this book. But the readers are advised to greet them with an open heart when they come their way.

## 7.5  Additional exercises

**Exercise 7.5.1.** Prove that if $X$ is a topological space, then the space $C_b(X)$ of all bounded continuous functions on $X$ with the supremum norm is a Banach space (replace $X$ with a metric space or even a subset of $\mathbb{R}^k$ if you are not comfortable with abstract topological spaces).

**Exercise 7.5.2.** Prove that a finite dimensional subspace of a normed space is closed, and that every finite dimensional normed space is a Banach space.

**Exercise 7.5.3.** Let $C^k([a, b])$ denote the space of functions (real-valued or complex-valued, whatever the reader prefers) that are $k$ times differentiable, such that $f, f', \ldots f^{(k)} \in C([a, b])$. For every $f \in C^k([a, b])$, define $\|f\|_{C^k} = \|f\|_\infty + \ldots + \|f^{(k)}\|_\infty$, that is,

$$\|f\|_{C^k} = \sup_{t \in [a,b]} |f(t)| + \sup_{t \in [a,b]} |f'(t)| + \ldots + \sup_{t \in [a,b]} |f^{(k)}(t)|.$$

Prove that $(C^k([a, b]), \|\cdot\|_{C^k})$ is a Banach space.

**Exercise 7.5.4.** For any $r, p \in [1, \infty]$ and every $m, n \in \mathbb{N}$, the space $M_{m,n}(\mathbb{C})$ of $m \times n$ matrices can be identified with the space of bounded operators $B(\ell_n^p, \ell_m^r)$. Find a formula, in terms of the coefficients of a matrix $A = (a_{ij})_{i,j=1}^{m,n}$, for the operator norms

$$\|A\|_{\ell_n^1 \to \ell_m^p} \quad \text{and} \quad \|A\|_{\ell_n^p \to \ell_m^\infty} \quad \text{for } p \in [1, \infty].$$

**Exercise 7.5.5.** Prove that an infinite matrix $A = (a_{ij})_{i,j=0}^\infty$ gives rise to a bounded operator on $\ell^1$ by matrix multiplication (similarly to Section 5.6), if and only if

$$\sup\left\{\sum_{i=0}^{\infty}|a_{ij}| : j \in \mathbb{N}\right\} < \infty.$$

Prove that the expression above is then equal to the operator norm $\|A\|$.

**Exercise 7.5.6.** Let $D$ be a nice, bounded subset of $\mathbb{R}^k$ (for example $D = [-R, R]^k$ for some $R > 0$, or any other set you know how to integrate on). Prove that for any $f, g \in C(D)$ the following continuous versions of Hölder's and Minkowski's inequalities hold:

$$\left|\int_D f(x)g(x)dx\right| \le \left(\int_D |f(x)|^p dx\right)^{1/p} \left(\int_D |g(x)|^q\right)^{1/q},$$

and

$$\left(\int_D |f(x) + g(x)|^p dx\right)^{1/p} \le \left(\int_D |f(x)|^p dx\right)^{1/p} + \left(\int_D |f(x)|^p dx\right)^{1/p}.$$

(**Hint:** Hölder's inequality follows from the discrete Hölder's inequality (Lemma 7.1.7) by first considering functions that attain finitely many values, and then using approximation considerations. If you get confused, prove for $k = 1$.)

**Exercise 7.5.7.** Let $X$ and $Y$ be normed spaces. Define a direct sum vector space

$$X \oplus Y = \{(x, y) : x \in X, y \in Y\},$$

with the obvious addition and multiplication by scalar operations. We introduce norms on $X \oplus Y$; for $p \in [1, \infty)$ we define

$$\|(x, y)\|_{\oplus, p} = (\|x\|_X^p + \|y\|_Y^p)^{1/p}.$$

We also define

$$\|(x, y)\|_{\oplus, \infty} = \max\{\|x\|_X, \|y\|_Y\}.$$

Prove that for every $p$, the space $(X \oplus Y, \|\cdot\|_{\oplus, p})$ is a Banach space. Prove that for every pair of conjugate exponents $p, q \in [1, \infty]$, the dual of $(X \oplus Y, \|\cdot\|_{\oplus, p})$ can be identified with $(X^* \oplus Y^*, \|\cdot\|_{\oplus, q})$; that is, loosely speaking:

$$(X \oplus Y, \|\cdot\|_{\oplus, p})^* = (X^* \oplus Y^*, \|\cdot\|_{\oplus, q}).$$

**Exercise 7.5.8.** The purpose of this exercise is to prove that normed spaces that have the same closures have the same completion.

Let $X$ be a normed space and let $Y \subset X$ be a dense subspace. Let $\widetilde{Y}$ and $\widetilde{X}$ be the completions of $Y$ and $X$, respectively. Let $V_Y : Y \to \widetilde{Y}$ and $V_X : X \to \widetilde{X}$ denote the embeddings. Prove that there is a surjective isometric isomorphism $U : \widetilde{Y} \to \widetilde{X}$ such that $UV_Y = V_X|_Y$. Moreover, show that if $X, Y$ are Hilbert spaces, then $U$ is a unitary. In other words, the completions of $X$ and $Y$ can be canonically identified via an isometric isomorphism that "fixes $Y$".

**Exercise 7.5.9** (Existence of best approximation in finite dimensional normed spaces). Let $X$ be a finite dimensional normed space, let $S \subseteq X$ be a closed convex subset, and let $x \in X$. Prove that there exists a point $s \in S$, such that

$$\|x - s\| = d(x, S) = \inf\{\|x - y\| : y \in S\}.$$

**Exercise 7.5.10** (Non-uniqueness of best approximation in Banach spaces). Give an example of a Banach space $X$, a closed convex subset $S \subseteq X$, and a point $x \in X$, such that there is more than one $s \in S$ for which

$$\|x - s\| = d(x, S) = \inf\{\|x - y\| : y \in S\}.$$

Does such an example exist if $S$ is assumed to be a linear subspace?

**Exercise 7.5.11** (Nonexistence of best approximation in Banach spaces). Construct an example of a Banach space $X$, a closed convex subset $S \subseteq X$, and an element $x \in X$, such that there is no $s \in S$ for which

$$\|x - s\| = d(x, S) = \inf\{\|x - y\| : y \in S\}.$$

(**Hint:** Let $X$ be $C_{\mathbb{R}}([0, 1])$ with the supremum norm, and let $S = \{f \in X : \int_0^1 f(t)dt = \frac{1}{2} \text{ and } f(0) = 1\}$.)

**Exercise 7.5.12** (Existence and uniqueness of best approximation in uniformly convex Banach spaces). A normed space $X$ is said to be **uniformly convex** if all $\epsilon > 0$, there exists $\delta > 0$, such that for all $x, y \in X_1$

$$\|x - y\| > \epsilon \text{ implies that } \left\|\frac{x+y}{2}\right\| < 1 - \delta.$$

1. Prove that every Hilbert space is uniformly convex.

2. Give an example of a Banach space, in which the norm is not induced by an inner product, that is uniformly convex.

3. Prove that in a uniformly convex Banach space $X$, for every closed and convex set $S \subseteq X$ and every $x \in X$, there exists a unique $s \in S$ such that
$$\|x - s\| = d(x, S) = \inf\{\|x - y\| : y \in S\}.$$

**Exercise 7.5.13.** Prove that, for all $p \in [1, \infty]$, the weak limit of a sequence in $\ell^p$ is unique.

**Exercise 7.5.14.** Let $p, q \in (1, \infty)$ be conjugate exponents, and let $(\ell^p)_1$ and $(\ell^q)_1$ be their unit balls. Let $\{g_k\}$ be a dense sequence in $(\ell^q)_1 = ((\ell^p)^*)_1$. Define $d : (\ell^p)_1 \times (\ell^p)_1 \to [0, \infty)$ by

$$d(x, y) = \sum_{k=1}^{\infty} 2^{-k}|g_k(x - y)|.$$

1. Prove that $d$ is a metric.

2. Prove that a sequence $\{x_n\}$ in $(\ell^p)_1$ converges weakly to $x \in (\ell^p)_1$ if and only if this series converges to $x$ in the metric $d$.

3. Prove that $((\ell^p)_1, d)$ is a compact metric space.

**Exercise 7.5.15** (Difficult). Prove that a sequence in $\ell^1$ is weakly convergent if and only if it is strongly convergent.

**Exercise 7.5.16.** Prove that a Hamel basis in an infinite dimensional Banach space must be uncountable (**Hint:** if $\{v_1, v_2, \ldots\}$ is a Hamel basis for $X$, then $X$ is the countable union $\cup_n \operatorname{span}\{v_1, \ldots, v_n\}$; now use Baire's theorem.)

**Exercise 7.5.17** (Principle of uniform boundedness). We called Theorem 7.3.11 a "baby version" of the principle of uniform boundedness. The general principle of uniform boundedness states that if $\{T_i\}_{i \in I}$ is a family of bounded operators from a Banach space $X$ into a Banach space $Y$, and if

$$\sup_{i \in I} \|T_i x\| < \infty \quad \text{for all } x \in X,$$

then

$$\sup_{i \in I} \|T_i\| < \infty.$$

In other words, a pointwise bounded family of operators is uniformly bounded. Prove the principle of uniform boundedness.

# Chapter 8

## The algebra of bounded operators on a Banach space

In the previous chapter we defined the space $B(X,Y)$ of bounded operators between two normed spaces. In this chapter, we will take a closer look at this space. We will make definitions and prove results in the general setting of operators acting between normed spaces, whenever we can do so easily. On the other hand, we will not hesitate to assume that the underlying spaces are Hilbert spaces, if this helps us obtain a stronger result or streamlines the presentation. The fact that a Hilbert space is its own dual (the Riesz representation theorem) makes the operator theory of Hilbert spaces far more tractable than the operator theory on any other infinite dimensional Banach space.

### 8.1 The algebra of bounded operators

Let $X, Y, Z$ be normed spaces. If $T : X \to Y$ and $S : Y \to Z$ are linear operators, then their composition $S \circ T$ is the map $S \circ T : X \to Z$ given by $S \circ T(x) = S(T(x))$. The composition $S \circ T$ is usually denoted by $ST$. We begin by making a simple and crucial observation.

**Proposition 8.1.1.** *Let $X, Y, Z$ be normed spaces. For every $S \in B(Y,Z)$ and $T \in B(X,Y)$, $ST \in B(X,Z)$ and in fact*

$$\|ST\| \le \|S\|\|T\|.$$

*Proof.* Immediate. □

We therefore obtain that for every normed space $X$, the normed space $B(X) = B(X,X)$ is an *algebra*, that is, $B(X)$ is a ring with unit, which is also a vector space (complex or real, depending on the scalar field of $X$), such that the vector space and ring operations are compatible.

If $T \in B(X)$, then $T^n = T \circ T \circ \cdots \circ T$ (the composition of $T$ with itself, $n$ times) is also in $B(X)$, and moreover $\|T^n\| \le \|T\|^n$. If $p(z) = \sum_{k=0}^{n} a_k z^k$ is a polynomial, then we can also naturally define $p(T)$ to be the operator

$$p(T) = \sum_{k=0}^{n} a_k T^k,$$

and it is clear that $p(T) \in B(X)$, because it is the sum of bounded operators. The "zeroth power" $T^0$ is understood as the identity operator: $T^0 = I$. When $c$ is a scalar, it is customary to write $c$ for the operator $cI$, thus we write

$$p(T) = a_0 + a_1 T + \cdots a_n T^n.$$

What can we say about the norm of $p(T)$? We can estimate the norm of $p(T)$ using the triangle inequality: $\|p(T)\| \leq \sum |a_k| \|T\|^k$. One can sometime do better; see Exercise 10.5.12.

---

## 8.2 An application to ergodic theory

### 8.2.1 The basic problem of ergodic theory

In the study of discrete dynamical systems, one considers the action of some map $T$ on a space $X$. *Ergodic theory* is the part of dynamical systems theory in which one is interested in the action of a measure preserving transformation $T$ on a measure space $X$. Perhaps surprisingly, the origins of ergodic theory are in mathematical physics - statistical mechanics, to be precise[1].

Since our goal here is merely to illustrate how operator theory can be applied to ergodic theory, we will work in the simplest possible setup: our space $X$ will be the unit interval $[0, 1]$, and our transformation $T : [0, 1] \to [0, 1]$ will be piecewise continuous (we will not, however, need to assume that $T$ is invertible). Anybody who took a course in measure theory will be able to generalize to the setting where $X$ is a probability space; the operator theoretic details will remain the same.

We assume further that $T$ is **measure preserving**. Recall from Example 7.2.10 that by this we mean that for all $f \in PC[0, 1]$,

$$\int_0^1 f \circ T(x) dx = \int_0^1 f(x) dx.$$

It is not entirely clear at first whether there are interesting examples of measure preserving maps.

**Example 8.2.1.** For $\alpha \in (0, 1)$, let $T(x) = x + \alpha \ (\mod 1)$, that is

$$T(x) = \begin{cases} x + \alpha & x + \alpha < 1 \\ x + \alpha - 1 & x + \alpha \geq 1 \end{cases}.$$

---

[1] See the section "Ergodic theory: an introduction" in [24].

Then

$$\int_0^1 f \circ T(x)dx = \int_0^{1-\alpha} f(x+\alpha)dx + \int_{1-\alpha}^1 f(x+\alpha-1)dx$$

$$= \int_\alpha^1 f(x)dx + \int_0^\alpha f(x)dx$$

$$= \int_0^1 f(x)dx.$$

**Example 8.2.2.** Let

$$T(x) = \begin{cases} 2x & x \in [0, 1/2] \\ 2x - 1 & x \in (1/2, 1] \end{cases}.$$

Then

$$\int_0^{1/2} f(2x)dx + \int_{1/2}^1 f(2x-1)dx = \frac{1}{2}\int_0^1 f(t)dt + \frac{1}{2}\int_0^1 f(t)dt$$

$$= \int_0^1 f(t)dt,$$

so $T$ is measure preserving. Note that $T([0, 1/2]) = [0, 1]$, so it does not do what one might naively think that a "measure preserving" map should do. However, $T$ does satisfy that the measure of $T^{-1}(A)$ is equal to the measure of $A$ for every $A \subseteq [0, 1]$, and this turns out to be the important property (the standard definition of a measure preserving transformation on $[0, 1]$, is a transformation $T$ such that the Lebesgue measure of $T^{-1}(A)$ is equal to the Lebesgue measure of $A$, for every measurable subset $A \subseteq [0, 1]$).

**Example 8.2.3.** One may also take $T(x) = 2x$ for $x \in [0, 1/2]$ and $T(x) = 2 - 2x$ for $x \in (1/2, 1]$, etc. It should be clear now how to define an infinite family of measure preserving transformations on $[0, 1]$.

Given some fixed measure preserving map $T$, we pick a point $x \in [0, 1]$, and we start moving it around the space $[0, 1]$ by applying $T$ again and again. We get a sequence $x, T(x), T^2(x), \ldots$ in $[0, 1]$. The basic problem in ergodic theory is to determine the statistical behavior of this sequence. To quantify the phrase "statistical behavior", one may study the large $N$ behavior of the so-called *time averages*

$$\frac{1}{N+1}\sum_{n=0}^N f(T^n(x)),$$

for functions $f \in PC[0, 1]$, say.

Why would this be interesting? Suppose, for example, that $f = \chi_{(a,b)}$ is the indicator function of some interval: $f(x) = 1$ if and only if $x \in (a, b)$, otherwise

$f(x) = 0$. In this case, the sum $\sum_{n=0}^{N} f(T^n(x))$ counts the number of times that $T^n(x)$ visited the interval $(a, b)$ in the first $N + 1$ steps that $x$ takes along the sequence $x, T(x), T^2(x), \ldots$. The time averages therefore measure the fraction of the "time" that the sequence $T^n(x)$ spends inside $(a, b)$. When one takes the limit $N \to \infty$, if that limit exists, one gets a measure of how much time the sequence spends inside $(a, b)$ in the long run.

If the sequence $T^n(x)$ behaves in a completely "random" manner, what would be our best guess for the limit of $\frac{1}{N+1} \sum_{n=0}^{N} f(T^n(x))$? If we think of the probability of $T^n(x)$ being at a certain point in $[0, 1]$ as being uniformly distributed on $[0, 1]$, then the best guess, intuitively, would be that

$$\lim_{N \to \infty} \frac{1}{N+1} \sum_{n=0}^{N} f(T^n(x)) = b - a.$$

Note that for $f = \chi_{(a,b)}$, the value $b - a$ is equal to $\int_0^1 f(t)dt$, so our guess is

$$\lim_{N \to \infty} \frac{1}{N+1} \sum_{n=0}^{N} f(T^n(x)) = \int_0^1 f(t)dt \qquad (8.1)$$

for indicator functions of intervals.

Now, maybe we would like to use some more complicated function $f$ to measure the distribution of the sequence $T^n(x)$. But if (8.1) holds for indicator functions then it holds for step functions, and then also presumably for other functions by some limiting process.

The equality (8.1) is called, sometimes, *the Ergodic Hypothesis*. It describes a situation where taking the **time average** (that is, starting at a point $x$ and taking the average of repeated measurements $f(T^n(x))$) is equal to the **space average** $\int_0^1 f(t)dt$ (which is the expected value of $f$ on the probability space $[0, 1]$). I certainly do not claim that we have justified (8.1); in fact (8.1) does not always hold. All that we said is that we might expect (8.1) to hold if the sequence $T^n(x)$ is spread out on the interval in a random or uniform way. The various ergodic theorems proved in ergodic theory make this very loose discussion precise.

## 8.2.2    The mean ergodic theorem

The mean ergodic theorem discusses the validity of (8.1) in the setting of $L^2[0, 1]$. (The reason why it is called *mean* is that convergence in the $L^2$ norm used to be called *convergence in the mean*). The first thing one has to do is to make sense of the composition $f \circ T$ in $L^2[0, 1]$. The problem is that if $f \in L^2[0, 1]$, then $f$ is not defined by the values it attains at points $x \in [0, 1]$, so it is not clear what we mean by $f \circ T$. However, we saw in Example 7.2.10 that a measure preserving transformation induces an isometric operator on $L^p[0, 1]$ for every $1 \le p < \infty$, given by

$$Uf = f \circ T$$

for $f \in C([0,1])$ and extended continuously to $L^p[0,1]$ for any $1 \le p < \infty$.

**Definition 8.2.4.** A transformation $T : [0,1] \to [0,1]$ is said to be ***ergodic*** if for all $f \in L^2[0,1]$, the equality $f = f \circ T$ implies that $f = const$. That is, $f$ is ergodic if the composition operator $U : L^2[0,1] \to L^2[0,1]$ has no fixed vectors, except the constant functions.

**Theorem 8.2.5** (von Neumann's mean ergodic theorem). *Let $T$ be a measure preserving (piecewise continuous) ergodic transformation on $[0,1]$. Then for all $f \in L^2[0,1]$*

$$\lim_{N \to \infty} \frac{1}{N+1} \sum_{n=0}^{N} f \circ T^n = \int_0^1 f(t)dt \tag{8.2}$$

*in the $L^2$ norm, where the right-hand side denotes the constant function with value $\int_0^1 f(t)dt$.*

**Remark 8.2.6.** Recalling the discussion surrounding (8.1), one immediately sees the limitation of this theorem. It does not tell us what happens for a particular $x$ in the time averages, but only gives us norm convergence of the sequence of functions $\frac{1}{N+1} \sum_{n=0}^{N} f \circ T^n$. Typically, pointwise (or almost everywhere pointwise) convergence theorems are harder to prove.

The operator $U$ that we defined above is an isometry, and in particular it satisfies $\|U\| \le 1$, that is, $U$ is a contraction. Theorem 8.2.5 is an immediate consequence of the following theorem.

**Theorem 8.2.7.** *Let $H$ be a Hilbert space and let $A \in B(H)$ be a contraction. Define $M = \{h : Ah = h\}$, and let $P_M$ be the orthogonal projection onto $M$. Then for all $h \in H$,*

$$\lim_{N \to \infty} \frac{1}{N+1} \sum_{n=0}^{N} A^n h = P_M h$$

*in norm.*

To see how Theorem 8.2.5 follows from Theorem 8.2.7, note simply that if $T$ is ergodic, then $M = \{f : Uf = f\}$ is the one-dimensional space of constant functions, so $P_M f = \langle f, 1 \rangle 1 = \int_0^1 f(t)dt$.

*Proof.* The first step of the proof requires a bit of inspiration. Recall that $\|A^*\| = \|A\| \le 1$. Let $h \in M$ be a unit vector, and consider $\langle A^*h, h \rangle = \langle h, Ah \rangle = \langle h, h \rangle = 1$. From the Cauchy-Schwarz inequality, $1 = |\langle A^*h, h \rangle| \le \|A^*h\| \|h\| \le 1$, so $|\langle A^*h, h \rangle| = \|A^*h\| \|h\|$. From the equality part of Cauchy-Schwarz this can only happen when $A^*h = ch$. But reconsidering $\langle A^*h, h \rangle = 1$ it is evident that $c = 1$. Thus $A^*h = h$. Because of the symmetry of the adjoint operation, we get a nice little result: *for a contraction $A$,*

$$Ah = h \quad \textit{if and only if} \quad A^*h = h.$$

Next, we try to understand what the decomposition $H = M \oplus M^{\perp}$ looks like. Note that $M = \ker(I - A)$. By our nice little result, $\ker(I - A) = \ker(I - A^*)$. Therefore, using Proposition 5.4.8, $M^{\perp} = \ker(I - A^*)^{\perp} = \overline{\mathrm{Im}(I - A)}$. So $M$ induces the decomposition

$$H = \ker(I - A) \oplus \overline{\mathrm{Im}(I - A)}.$$

If $h \in M = \ker(I - A)$, then $\frac{1}{N+1} \sum_{n=0}^{N} A^n h = h = P_M h$, and the conclusion of the theorem holds for this $h$.

Let $g \in \mathrm{Im}(I - A)$. Then $g$ has the form $g = f - Af$ for some $f$. We compute

$$\frac{1}{N+1} \sum_{n=0}^{N} A^n (f - Af) = \frac{1}{N+1}(f - A^{N+1}f) \to 0.$$

We have used the fact that $\|A^{N+1}\| \le \|A\|^{N+1}$, so that $f - A^{N+1}f$ is bounded.

Now consider $y \in \overline{\mathrm{Im}(I - A)}$, and fix $\epsilon > 0$. There is some $g \in \mathrm{Im}(I - A)$ such that $\|y - g\| < \epsilon$. For some $N_0$, $\|\frac{1}{N+1} \sum_{n=0}^{N} A^n g\| < \epsilon$ for all $N \ge N_0$. Then, for $N \ge N_0$ we break up $\sum_{n=0}^{N} A^n y$ as $\sum_{n=0}^{N} A^n (y - g) + \sum_{n=0}^{N} A^n g$ to find that

$$\left\| \frac{1}{N+1} \sum_{n=0}^{N} A^n y \right\| \le \left\| \frac{1}{N+1} \sum_{n=0}^{N} A^n \right\| \|y - g\| + \left\| \frac{1}{N+1} \sum_{n=0}^{N} A^n g \right\|$$

$$< 2\epsilon,$$

demonstrating that $\frac{1}{N+1} \sum_{n=0}^{N} A^n y \to 0 = P_M y$. Thus the theorem holds for all $y \in M^{\perp}$. Finally, since $H = M \oplus M^{\perp}$, it follows that the theorem holds on all of $H$. $\qquad\square$

## 8.3    Invertible operators and inverses

**Definition 8.3.1.** An operator $A \in B(X, Y)$ is said to be *invertible* if there exists $B \in B(Y, X)$ such that

$$AB = I_Y , \quad BA = I_X.$$

In this case $B$ is called the *inverse* of $A$, and is denoted $A^{-1}$.

Thus, in the case $X = Y$, $A$ is said to be invertible if it is an invertible element of the algebra $B(X)$. It is clear that if $A$ is invertible, then it is bijective. What may not be clear — and rightly so, because it is a nontrivial question — is whether or not a bounded linear operator which is bijective, is also invertible in the sense of this definition. To emphasize, a bijective linear operator $A$ always has a linear inverse $A^{-1}$. The question is, if in addition $A$ is assumed bounded, does it follow that $A^{-1}$ is also bounded?

This turns out to be true when $X$ and $Y$ are Banach spaces, and this fact is known as the *inverse mapping theorem*. It is a special case of the *open mapping theorem*, the general case of which we shall not discuss. We will prove the inverse mapping theorem for operators on Hilbert spaces.

**Definition 8.3.2.** An operator $A \in B(X, Y)$ is said to be **bounded below** if there exists $\epsilon > 0$ such that

$$\|Ax\| \geq \epsilon \|x\|, \quad \text{for all } x \in X.$$

**Exercise 8.3.3.** Let $X$ and $Y$ be Banach spaces and $A \in B(X, Y)$.

1. If $A$ is bounded below, then for every closed subspace $F \subseteq X$, $A(F)$ is closed.

2. If $A$ is bounded below and $\text{Im}(A)$ is dense, then $A$ is invertible.

## 8.3.1    The inverse mapping theorem in Hilbert spaces

Our proof of the inverse mapping theorem relies on the following neat fact (of independent interest), which should be considered as a kind of converse to the fact that every bounded operator between Hilbert spaces has an adjoint.

Throughout this section, $H$ and $K$ denote Hilbert spaces.

**Theorem 8.3.4.** *Let $A : H \to K$ and $B : K \to H$ be linear maps. If*

$$\langle Ah, k \rangle = \langle h, Bk \rangle,$$

*for all $h \in H$ and $k \in K$, then $A$ and $B$ are bounded, and $B = A^*$.*

*Proof.* It suffices to show that $A$ is bounded, the rest follows easily. By Exercise 7.3.13, it suffices to show that $A$ is *weakly sequentially continuous*, in the sense that

$$h_n \rightharpoonup h \quad \text{implies that} \quad Ah_n \rightharpoonup Ah.$$

But if $h_n \rightharpoonup h$, then for all $k \in K$,

$$\langle Ah_n, k \rangle = \langle h_n, Bk \rangle \to \langle h, Bk \rangle = \langle Ah, k \rangle,$$

showing that $Ah_n \rightharpoonup Ah$, and we are done. $\square$

**Theorem 8.3.5** (Inverse mapping theorem). *Let $A \in B(H, K)$ be a bijective linear operator. Then $A$ has a bounded inverse.*

*Proof.* Let $A^{-1}$ be the (algebraic) inverse of $A$. We need to show that $A^{-1}$ is bounded. By Theorem 8.3.4, it suffices to show that there exists a linear map $B : H \to K$ such that for all $h \in H$, $k \in K$,

$$\langle A^{-1}k, h \rangle = \langle k, Bh \rangle.$$

We will show that $A^*$ is (algebraically) invertible. It is then clear that $B = (A^*)^{-1}$ does the trick; indeed, just consider

$$\langle k, (A^*)^{-1}h \rangle = \langle AA^{-1}k, (A^*)^{-1}h \rangle = \langle A^{-1}k, h \rangle.$$

Now, $\ker A^* = \operatorname{Im} A^\perp = 0$, so $A^*$ is one-one. Moreover, $\overline{\operatorname{Im} A^*} = \ker A^\perp = H$, so $A^*$ has dense image. It is left to show that $\operatorname{Im} A^*$ is closed.

Assume that $h \in \overline{\operatorname{Im} A^*}$, so $A^*k_n \to h$ for some sequence $\{k_n\}$ in $K$. For any $g \in K$

$$\langle k_n, g \rangle = \langle k_n, AA^{-1}g \rangle = \langle A^*k_n, A^{-1}g \rangle,$$

and the right-hand side converges to $\langle h, A^{-1}g \rangle$. This means that $\{k_n\}$ is a weakly convergent sequence (see Exercise 7.3.14), so there is a $y \in K$ such that $k_n \rightharpoonup y$.

But since $A^*$ is continuous, it follows from Exercise 7.3.13 that $A^*k_n \rightharpoonup A^*y$. On the other hand, $A^*k_n \to h$. From uniqueness of the weak limit we have that $h = A^*y \in \operatorname{Im} A^*$, and this completes the proof. $\square$

**Exercise 8.3.6.** Let $H, K$ be Hilbert spaces and let $A \in B(H, K)$. The following are equivalent.

1. $A$ is invertible.

2. $A$ is bounded below and $\operatorname{Im} A$ is dense in $K$.

3. $A$ and $A^*$ are bounded below.

4. $A^*$ is invertible.

### 8.3.2 Perturbations of invertible operators

We return now to the study of bounded operators on general Banach spaces. Our goal in this section is to prove that an operator which is "sufficiently close" to an invertible operator is also invertible. Before proving the basic result, we need to discuss convergence of series in normed spaces.

**Definition 8.3.7.** Let $\{x_n\}_{n=1}^\infty$ be a sequence in a normed space $X$. We say the series $\sum_{n=1}^\infty x_n$ *converges absolutely* if

$$\sum_{n=1}^\infty \|x_n\| < \infty.$$

It is easy to see that if $\sum x_n$ converges absolutely, then the sequence $\left\{\sum_{n=1}^{N} x_n\right\}_{N=1}^{\infty}$ is a Cauchy sequence. Thus, if $X$ is complete, then there exists a limit

$$\sum_{n=1}^{\infty} x_n = \lim_{N \to \infty} \sum_{n=1}^{N} x_n.$$

It is also clear that this limit does not depend on the order in which the sum is taken. Therefore, this sum also exists in the same sense as when discussing convergence of infinite series in Hilbert spaces. Making the obvious modification of Definition 3.3.4 to normed spaces, we can say that *in a Banach space, an absolutely convergent series is also convergent.*

We also have the following fact, which is sometimes useful.

**Exercise 8.3.8.** Let $X$ be a normed space. The following are equivalent:

1. $X$ is complete.

2. Every absolutely convergent series converges to an element of $X$.

From here on, let $X$ be a Banach space.

**Proposition 8.3.9.** *Let $A \in B(X)$, $\|A\| < 1$. Then $I - A$ is invertible, and*

$$(I - A)^{-1} = \sum_{n=0}^{\infty} A^n. \tag{8.3}$$

**Remark 8.3.10.** The formula $(8.3)$ for the inverse of $I - A$ is referred to as the *Neumann series* of $A$. The reader must have noticed the analogy with the Taylor series of $(1 - z)^{-1}$ for $|z| < 1$. There is a way to give meaning to the expression $f(A)$ when $f$ is an analytic function; see Exercise 8.5.11.

*Proof.* Since $\|A^n\| \le \|A\|^n$ and $\|A\| < 1$, we have that the Neumann series of $A$ converges absolutely, thus (since $B(X)$ is complete) there is a well-defined element

$$\sum_{n=0}^{\infty} A^n = \lim_{N \to \infty} \sum_{n=0}^{N} A^n \in B(X).$$

We have to check that $(I - A) \sum A^n = \sum A^n (I - A) = I$. Now,

$$(I - A) \sum_{n=0}^{\infty} A^n = \lim_{N \to \infty} (I - A) \sum_{n=0}^{N} A^n,$$

(why?), and

$$(I - A) \sum_{n=0}^{N} A^n = I - A^{N+1} \to I,$$

so $(I - A) \sum A^n = I$. Similarly, $\sum A^n (I - A) = I$. $\qquad\square$

**Corollary 8.3.11.** *Let $A \in B(X)$ be invertible. For every $B \in B(X)$, if $\|A - B\| < \frac{1}{\|A^{-1}\|}$, then $B$ is also invertible.*

**Exercise 8.3.12.** Prove the above corollary. (**Hint:** write $I - A^{-1}B = A^{-1}(A - B)$ and use Proposition 8.3.9 to write down an inverse for $A^{-1}B$.)

**Definition 8.3.13.** Let $GL(X)$ denote the set of invertible elements in $B(X)$. $GL(X)$ is referred to as **the general linear group** of $X$.

**Exercise 8.3.14.** By Corollary 8.3.11, $GL(X)$ is an open subset of $B(X)$. Show that when endowed with the multiplication and topology of $B(X)$, $GL(X)$ is a *topological group*, meaning that it is a group in which the group operations (multiplication and inverse) are continuous. (**Hint:** to show that $A \mapsto A^{-1}$ is continuous, examine the form of the inverse obtained in the previous exercise.)

### 8.3.3    The spectrum of a bounded operator

In this section, $X$ denotes a complex Banach space.

**Definition 8.3.15.** Let $A \in B(X)$. The **resolvent set** of $A$ is the set

$$\rho(A) = \{\lambda \in \mathbb{C} : A - \lambda I \in GL(X)\}.$$

The **spectrum** of $A$, $\sigma(A)$, is defined to be the complement of the resolvent set, that is

$$\sigma(A) = \mathbb{C} \setminus \rho(A) = \{\lambda \in \mathbb{C} : A - \lambda I \notin GL(X)\}.$$

**Definition 8.3.16.** Let $A \in B(X)$. If $\lambda \in \mathbb{C}$ is a scalar for which there exists a nonzero $x \in X$ such that $Ax = \lambda x$, then $\lambda$ is said to be an **eigenvalue of** $A$, and $x$ is said to be an **eigenvector** corresponding to $\lambda$. The space spanned by all eigenvectors corresponding to an eigenvalue $\lambda$ is called the **eigenspace** corresponding to $\lambda$. The **point spectrum** of $A$, denoted by $\sigma_p(A)$ is the set of all eigenvalues of $A$.

In other words, $\lambda$ is an eigenvector if and only if $A - \lambda I$ is not injective. If $X$ is finite dimensional, $\sigma(A) = \sigma_p(A)$. However, when $X$ is infinite dimensional, then the spectrum of an operator can contain points that are not eigenvalues.

**Example 8.3.17.** Let $\{w_n\}$ be a sequence of positive numbers tending to 0. Define an operator $A : \ell^2 \to \ell^2$ by

$$A(a_n) = (w_n a_n).$$

Then $A$ is a bounded operator, and $\|A\| = \sup |w_n|$. Moreover, from the definitions we see that $\lambda \in \sigma_p(A)$ if and only if $\lambda = w_n$ for some $n$. In particular, 0 is not an eigenvalue. On the other hand, $A = A - 0I$ is not invertible, because if it was, it would have to be bounded below (Exercise 8.3.6). Thus 0 is a point in the spectrum which is not an eigenvalue.

In fact, there are operators with no eigenvalues whatsoever. On the other hand, the spectrum is never empty.

**Theorem 8.3.18.** *For every $A \in B(X)$, the spectrum $\sigma(A)$ is a nonempty, compact subset of $\mathbb{C}$ contained in $\{z \in \mathbb{C} : |z| \leq \|A\|\}$.*

*Proof.* We shall not prove[2] that $\sigma(A)$ is nonempty; this beautiful proof will be given in a more advanced course in functional analysis (we will, however, prove later that the spectrum is nonempty for self-adjoint operators on a Hilbert space, and also for compact operators on a Banach space).

Corollary 8.3.11 shows that $\rho(A)$ is open, thus $\sigma(A)$ is closed. Now if $|\lambda| > \|A\|$, then $\|\lambda^{-1}A\| < 1$, so $\lambda^{-1}A - I$ is invertible, hence $A - \lambda I$ is, too. Thus, $\sigma(A)$ is contained in $\{z \in \mathbb{C} : |z| \leq \|A\|\}$. $\quad\square$

**Remark 8.3.19.** If we were working over the reals, then it would not be true that the spectrum is nonempty. This is one of the most significant differences between operator theory over the reals and operator theory over the complex numbers. Recall that in finite dimensional linear algebra, the fact that every operator has an eigenvalue follows from the fundamental theorem of algebra. In infinite dimensional spaces, the proof that $\sigma(A)$ is nonempty uses complex function theory in a nontrivial way.

**Exercise 8.3.20.** For an operator $A$ on a Hilbert space, show that

$$\sigma(A^*) = \{\bar{\lambda} : \lambda \in \sigma(A)\}.$$

**Exercise 8.3.21.** Let $S$ be the unilateral shift on $\ell^2$, defined in Example 5.5.8. Compute $\sigma(S)$ and $\sigma_p(S)$ (**Hint:** consider also $S^*$.)

---

## 8.4   *Isomorphisms

We pause the development of the theory to discuss the question *when are two Banach spaces the same?* Recall that Hilbert spaces which had orthonormal bases of the same cardinality turned out to be isomorphic, so they all enjoy the same structure and the same geometry. When it comes to Banach spaces the landscape is much more diverse. Although we hardly have any tools to begin mapping the landscape at this point, we raise the issue now, so we can think about this interesting problem as we go along.

---

[2]This is the only result in this book that is stated without proof, the reason being that any proof of it requires a theorem from complex analysis (typically Liouville's theorem), which the student is not assumed to know. Accordingly, this result will not be used in the sequel.

### 8.4.1 Isomorphisms and equivalent norms

In the following discussion, let $X$ and $Y$ be normed spaces. Recall that, by Definition 8.3.1, a bounded operator $T$ is said to be invertible if it is bijective and has a bounded inverse.

**Definition 8.4.1.** An invertible operator $T \in B(X,Y)$ is called an *isomorphism*. The spaces $X$ and $Y$ are said to be *isomorphic* if there is an isomorphism between them.

**Definition 8.4.2.** An operator $T \in B(X,Y)$ is said to be an *isometry* if $\|Tx\| = \|x\|$ for all $x \in X$. The spaces $X$ and $Y$ are said to be *isometrically isomorphic* (or simply *isometric*) if there is an isometric isomorphism between them.

Note that the notion of isomorphism that we defined here is much weaker than the one we defined for Hilbert spaces in Definition 3.4.3 (though no confusion should occur, as Exercise 8.4.3 below shows). Indeed, a Hilbert space cannot be isometrically isomorphic to a Banach space that is not Hilbert (why?), but it certainly can be isomorphic to one. In fact, we will soon see that all complex $n$-dimensional normed spaces are isomorphic to $\mathbb{C}^n$ with the standard inner product.

**Exercise 8.4.3.** Let $H$ and $K$ be Hilbert spaces.

1. Prove that if $H$ is isomorphic to $K$ as Banach spaces, then they are also isomorphic as Hilbert spaces in the sense of Definition 3.4.3.

2. Give an example of a Banach space isomorphism between Hilbert spaces that is not a Hilbert space isomorphism.

3. Prove that if $T : H \to K$ is an isometric isomorphism, then it is a unitary map.

**Exercise 8.4.4.** Prove that $\ell^p$ is not isomorphic to $\ell^\infty$ for all $1 \leq p < \infty$. (Harder: what can you say if we replace $\ell^\infty$ by $\ell^q$, for some $q \in [1, \infty)$?)

**Definition 8.4.5.** Let $X$ be a normed space with a norm $\|\cdot\|$, and let $\|\cdot\|'$ be another norm defined on $X$. The norms $\|\cdot\|$ and $\|\cdot\|'$ are said to be *equivalent* if there exists two positive constants $c, C$ such that

$$c\|x\| \leq \|x\|' \leq C\|x\| \quad \text{for all } x \in X.$$

Clearly $\|\cdot\|$ and $\|\cdot\|'$ are equivalent if and only if the identity map is an isomorphism from $(X, \|\cdot\|)$ onto $(X, \|\cdot\|')$.

**Exercise 8.4.6.** Let $(X, \|\cdot\|_X)$ and $(Y, \|\cdot\|_Y)$ be normed spaces, and let $T : X \to Y$ be a bijective linear map. Prove that if we define $\|x\|' = \|Tx\|_Y$, then $\|\cdot\|'$ is a norm on $X$. Prove that $T$ is an isomorphism from $(X, \|\cdot\|_X)$ onto $(Y, \|\cdot\|_Y)$ if and only if $\|\cdot\|$ and $\|\cdot\|'$ are equivalent.

**Exercise 8.4.7.** Prove that the $\|\cdot\|_p$ norms on $\ell_0$ (the space of all finitely supported sequences) are all mutually nonequivalent.

## 8.4.2   Finite dimensional normed spaces

**Theorem 8.4.8.** *All norms on $\mathbb{C}^n$ are equivalent. In other words, every two normed spaces of the same dimension over $\mathbb{C}$ are isomorphic.*

**Remark 8.4.9.** This same result — with the same proof — holds true for real spaces.

*Proof.* Let $\ell_n^2$ denote the space $\mathbb{C}^n$ with the norm $\|\cdot\|_2$. Let $X$ be an $n$-dimensional vector space with norm $\|\cdot\|$. We know from linear algebra that every $n$-dimensional vector space over $\mathbb{C}$ is linearly isomorphic to $\mathbb{C}^n$, so let $T : \ell_n^2 \to X$ be a bijective linear map. We need to prove that $T$ is bounded and bounded below (Exercise 8.3.3).

Let $e_1, \ldots, e_n$ be the standard basis of $\ell_n^2$. Put

$$M = \left( \sum_{k=1}^n \|T(e_k)\|^2 \right)^{1/2}.$$

Then

$$\left\| T\left( \sum_{k=1}^n a_k e_k \right) \right\| \leq \sum_{k=1}^n |a_k| \|T(e_k)\| \leq M \left( \sum_{k=1}^n |a_k|^2 \right)^{1/2} = M \left\| \sum_{k=1}^n a_k e_k \right\|_2,$$

where the second inequality follows from the Cauchy-Schwarz inequality. This shows that $T : \ell_n^2 \to X$ is a bounded operator with $\|T\| \leq M$. On the other hand, for every $x \neq 0$, $Tx \neq 0$. Therefore, since $T$ is continuous and $\{x \in \mathbb{C}^n : \|x\|_2 = 1\}$ is compact,

$$\inf_{\|x\|_2 = 1} \|Tx\| > 0.$$

It follows that $T$ is bounded below, and the proof is complete. $\square$

## 8.5   Additional exercises

**Exercise 8.5.1.** Let $T : [0,1] \to [0,1]$ be a piecewise continuous map.

1. Prove the converse of the mean ergodic theorem: *if for every $f \in L^2[0,1]$,*

$$\lim_{N \to \infty} \frac{1}{N+1} \sum_{n=0}^N f \circ T^n = \int_0^1 f(t)dt$$

*in the $L^2$ norm, then $T$ is ergodic.*

2. Prove that the transformation $T(x) = x + \alpha \ (\mod 1)$ is ergodic if and only if $\alpha$ is an irrational number.

3. Prove that if $\alpha$ is an irrational number and $T$ is as above, then, for every $f \in C([0,1])$, and for *every point* $x \in [0,1]$, the time averages $\frac{1}{N+1} \sum_{n=0}^{N} f(T^n(x))$ converge, as $N \to \infty$, to the mean $\int_0^1 f(t)dt$.

**Exercise 8.5.2.** Let $A = (a_{ij})_{i,j=0}^{\infty}$ be an infinite matrix where $a_{ij} = 2^{-i-j-1}$. Prove that $A$ defines a bounded operator on $\ell^p$ (given by matrix multiplication) for $p = 1, 2, \infty$. For which of these values of $p$ is this operator invertible?

**Exercise 8.5.3** (Closed graph theorem for Hilbert spaces). Let $H$ and $K$ be Hilbert spaces, and let $T : H \to K$ be a linear operator. The **graph of** $T$ is the following linear subspace of $H \oplus K$:

$$G(T) = \{(h, Th) : h \in H\}.$$

The operator $T$ is said to have a **closed graph** if $G(T)$ is a closed subspace of the Hilbert space $H \oplus K$.

1. Prove that if $T$ is bounded, then $G(T)$ is closed.

2. Prove that $G(T)$ is closed if and only if the following condition holds: *whenever there is a sequence $\{h_n\} \subset H$ such that $h_n \to h$ and $Th_n \to k$, then $k = Th$.*

3. Prove that if $G(T)$ is closed, then $T$ is bounded. (**Hint:** consider the bijective linear operator $h \mapsto (h, Th)$ of $H$ onto $G(T)$, and use Theorem 8.3.5.)

The statement that an operator with closed graph is bounded is called the *closed graph theorem*. It also holds true for operators between Banach spaces, and the proof is the same; one just needs the *inverse mapping theorem* for general Banach spaces.

**Exercise 8.5.4.** Let $H$ be a Hilbert function space on a set $X$, and let $f : X \to \mathbb{C}$. Suppose that for every $h \in H$, the function $fh$ is also in $H$. Prove that the multiplication operator

$$M_f : h \mapsto fh, \quad h \in H$$

is bounded.

**Exercise 8.5.5.** Let $A = (a_{ij})_{i,j=0}^{\infty}$ be an infinite matrix, and assume that every row of $A$ is in $\ell^2$, i.e., $\sum_{j=0}^{\infty} |a_{ij}|^2 < \infty$ for all $i \in \mathbb{N}$. Assume further that matrix multiplication by $A$ gives rise to a well-defined linear operator, which is also denoted by $A$. Prove that $A$ has to be bounded.

**Exercise 8.5.6** (Open mapping theorem in Hilbert spaces). Let $X$ and $Y$ be Banach spaces. A map $T : X \to Y$ is said to be an **open map** if for every open set $U \subseteq X$, the image $T(U)$ is an open set in $Y$.

1. Prove that $T$ is an open map if and only if there exists a positive $r > 0$, such that
$$T\left(\{x \in X : \|x\| < 1\}\right) \supset \{y \in Y : \|y\| < r\}.$$

2. Prove that if $T : H \to K$ is a surjective and bounded operator between Hilbert spaces, then it is an open map. (**Hint:** consider $T\big|_{\ker T^\perp} \cdot$)

It is also true in the setting of Banach spaces that a surjective and bounded linear map is an open map — this is the *open mapping theorem*. We challenge the reader to try to prove this general result (the proof does not require learning new material, only a bit of ingenuity).

**Exercise 8.5.7.** Let $X$ be a Banach space and let $T \in B(X)$. Assume that there is an $N \in \mathbb{N}$ for which $\|T^N\| < 1$. Prove or give a counterexample to each of the following claims:

1. $\|T\| < 1$.

2. $T^n \xrightarrow{n \to \infty} 0$.

3. $I - T$ is invertible.

**Exercise 8.5.8.** Let $a \in C_{per}([0, 1])$ such that $\|a\|_\infty \leq 1$. Let $T : [0, 1] \to [0, 1]$ be the piecewise continuous map $T(x) = x + a(\mod 1)$. Suppose that $\{x \in [0, 1] : |a(x)| < 1\} \neq \emptyset$. Prove that for every $g \in C_{per}([0, 1])$, there exists a unique $f \in C_{per}([0, 1])$ that satisfies the following *functional equation*:
$$f(x) + a(x)f(T(x)) = g(x) \quad \text{for all } x \in [0, 1].$$

Write down a formula for the solution $f$ in terms of $a, T$ and $g$.

**Exercise 8.5.9.** Let $c$ and $c_0$ be the subspaces of $\ell^\infty$ given by
$$c = \{x = (x_n)_{n=0}^\infty \in \ell^\infty : \lim_{n \to \infty} x_n \text{ exists}\},$$
and
$$c_0 = \{x = (x_n)_{n=0}^\infty \in \ell^\infty : \lim_{n \to \infty} x_n = 0\}.$$

1. Prove that $(c_0)^*$ is isometrically isomorphic to $\ell^1$.

2. Prove that $c^*$ is also isometrically isomorphic to $\ell^1$.

3. Prove that $c_0$ is isomorphic as a Banach space to $c$.

4. Prove that $c_0$ is not isometrically isomorphic to $c$ (no hint; this is meant to challenge your creativity).

**Exercise 8.5.10.** Recall that a bounded operator $T$ on a complex Hilbert space $H$ is said to be *positive* if $\langle Th, h \rangle \geq 0$ for all $h \in H$. An operator $T$ is said to be *strictly positive* if there exists a positive constant $c$ such that $\langle Th, h \rangle \geq c\|h\|^2$ for all $h \in H$. Prove or give a counterexample to each of the following claims:

1. $T$ is strictly positive if and only if $T$ is positive and invertible.

2. $T$ is strictly positive if and only if $\langle Th, h \rangle > 0$ for all $h \in H$.

3. Let $A$ and $B$ be two positive operators on $H$, let $M$ be a closed subspace of $H$, and let $P = P_M$ and $Q = I - P = P_{M^\perp}$. If $PAP$ is strictly positive on $M$ and $QBQ$ is strictly positive on $M^\perp$, then $A + B$ is strictly positive on $H$.

4. Let $A$ and $B$ be two positive operators on $H$, let $M$ be a closed subspace of $H$, and let $P = P_M$ and $Q = I - P = P_{M^\perp}$. Now we add the assumption that $M$ is invariant under $A$ (i.e., $AM \subseteq M$). If $PAP$ is strictly positive on $M$ and $QBQ$ is strictly positive on $M^\perp$, then $A + B$ is strictly positive on $H$.

**Exercise 8.5.11** (Analytic functional calculus). Let $D_r(0) = \{z \in \mathbb{C} : |z| < r\}$ be the disc of radius $r > 0$ around 0. We let $\mathcal{O}(D_r(0))$ denote the set of all analytic (or holomorphic) functions on $D_r(0)$. Recall that $g \in \mathcal{O}(D_r(0))$ if and only if $g$ has a power series representation $g(z) = \sum_{n=0}^{\infty} a_n z^n$, that converges absolutely in $D_r(0)$. The radius of convergence $R$ of the power series is given by $R^{-1} = \limsup |a_n|^{1/n}$, and satisfies $R \geq r$.

Let $X$ be a Banach space.

1. Let $A \in B(X)$ such that $\|A\| < r$. Prove that if $g(z) = \sum a_n z^n \in \mathcal{O}(D_r(0))$, the series $\sum_{n=0}^{\infty} a_n A^n$ converges in norm. Define $g(A) = \sum_{n=0}^{\infty} a_n A^n$.

2. Prove that if $f \in \mathcal{O}(D_r(0))$ too, then $(f+g)(A) = f(A) + g(A)$ and that $fg(A) = f(A)g(A)$.

3. Prove that if $\|A_n\| < r$ for all $n$ and if $A_n \to A$, then $g(A_n) \to g(A)$ (**Hint:** use

$$A^n - B^n = (A - B)(A^{n-1} + A^{n-2}B + \dots AB^{n-2} + B^{n-1})$$

and the radius of convergence.)

4. Suppose that $h \in \mathcal{O}(D_\rho(0))$ and that $\|g(A)\| < \rho$. Prove that $h \circ g(A) = h(g(A))$.

Given $A \in B(X)$, the mapping $g \mapsto g(A)$ is called **analytic** (or **holomorphic**) **functional calculus**. In fact, the analytic functional calculus can be extended to a larger class of functions and operators: whenever $U \subseteq \mathbb{C}$ is an open set (not necessarily a disc) and $g \in \mathcal{O}(U)$ (not necessarily given by a power series that converges in all of $U$), one can define $g(A)$ for every $A \in B(X)$ with $\sigma(A) \subset U$; for details, see, e.g., [16, Section 3.3].

**Exercise 8.5.12.** Following Exercise 8.5.11, we define for all $A \in B(X)$

$$\exp(A) = \sum_{n=0}^{\infty} \frac{A^n}{n!}.$$

1. Prove that if $A$ and $B$ commute (i.e., $AB = BA$), then

$$\exp(A + B) = \exp(A)\exp(B)$$

2. Prove that if $A$ is a self-adjoint element of $B(H)$ ($H$ a Hilbert space), then $\exp(iA)$ is a unitary operator.

3. Fix $A \in B(X)$, and define for every $t \in [0, \infty)$

$$T_t = \exp(tA).$$

Show that $\{T_t\}_{t \geq 0}$ forms a *uniformly continuous semigroup*, in the sense that **(i)** $T_0 = I_X$; **(ii)** $T_{s+t} = T_s T_t$ for all $s, t \geq 0$; and **(iii)** $\lim_{t \to t_0} T_t = T_{t_0}$.

4. Prove that the semigroup $\{T_t\}_{t \geq 0}$ is differentiable at 0, in the sense that the limit

$$\lim_{t \searrow 0} \frac{T_t - I}{t}$$

exists. What is the value of this limit?

**Exercise 8.5.13.** For every $t \in [0, \infty)$, we define an operator $S_t$ on $C([0,1])$ by

$$(S_t f)(x) = \begin{cases} 0 & x \in [0, t) \\ f(x - t) & x \in [0,1] \cap [t, \infty) \end{cases}.$$

1. Prove that $S_t$ is bounded on $(C([0,1]), \|\cdot\|_1)$, where $\|f\|_1 = \int_0^1 |f(t)| dt$. Therefore $S_t$ extends to a bounded operator on $L^1[0,1]$ (the completion of $(C([0,1]), \|\cdot\|_1)$).

2. Prove that the family $\{S_t\}_{t \geq 0}$ forms a *strongly continuous semigroup*, in the sense that **(i)** $S_0 = I$; **(ii)** $S_{s+t} = S_s S_t$ for all $s, t \geq 0$; and **(iii)** $\lim_{t \to t_0} S_t f = S_{t_0} f$ for all $f \in L^1[0,1]$.

3. Prove that for every $f$ in the subspace $C^1([0,1])$ of continuously differentiable functions on $[0,1]$, the limit

$$\lim_{t \searrow 0} \frac{S_t f - f}{t}$$

exists in $L^1[0,1]$ and defines a linear operator $A : C^1([0,1]) \to L^1[0,1]$. Describe the operator $A$.

4. Is the operator $A$ bounded on $(C^1([0,1]), \|\cdot\|_1)$?

5. Prove that the limit

$$\lim_{t \searrow 0} \frac{S_t - I}{t}$$

does not exist in the operator norm.

**Exercise 8.5.14.** Let $U$ be the bilateral shift on $\ell^2(\mathbb{Z})$ (see Exercise 5.7.3). For each of the operators $U$ and $U^*$ find the spectrum and the point spectrum.

**Exercise 8.5.15.** Let $T$ be a linear operator and let $v_1, \ldots, v_n$ be $n$ eigenvectors corresponding to $n$ distinct eigenvalues $\lambda_1, \ldots, \lambda_n$. Prove that $v_1, \ldots, v_n$ are linearly independent.

**Exercise 8.5.16.** For every compact set $K$ in the complex plane, prove that there exists a bounded operator $T$ on a separable Hilbert space such that $\sigma(T) = K$. Much harder: does there also exist a bounded operator on a separable Hilbert space such that $\sigma_p(T) = K$?

**Exercise 8.5.17.** Let $A$ be a self-adjoint operator on a Hilbert space $H$. Define an operator $T$ on $H \oplus H$ by

$$T = \begin{pmatrix} 0 & A \\ A & 0 \end{pmatrix}.$$

Find $\sigma(T)$ and $\sigma_p(T)$ in terms of $\sigma(A)$ and $\sigma_p(A)$.

**Exercise 8.5.18.** On the space $C^1([a, b])$ we define three norms:

1. $\|f\| = \|f\|_\infty + \|f'\|_\infty$ (recall Exercise 7.5.3),

2. $\|f\|' = |f(0)| + \|f'\|_\infty$,

3. $\|f\|^* = \|f\|_\infty + |f'(0)|$.

Determine which of these norms is equivalent to one of the others.

# Chapter 9

## Compact operators

### 9.1 Compact operators

Throughout, $X$ and $Y$ denote Banach spaces. Recall that the closed unit ball in $X$ is oftentimes denoted $X_1$, or $(X)_1$.

**Definition 9.1.1.** A linear operator $A : X \to Y$ is said to be **compact** if the image of $X_1$ under $A$ is precompact (i.e., $\overline{A(X_1)}$ is compact in $Y$). We let $K(X, Y)$ denote the space of all compact operators from $X$ to $Y$, and we write $K(X)$ for $K(X, X)$.

Since a subset of a metric space is compact if and only if it is sequentially compact (Theorem A.3.27), we have the following convenient reformulation of the definition: *An operator $A$ is compact if and only if for every bounded sequence $\{x_n\}$ in $X$, there is a subsequence $\{x_{n_k}\}$ such that $\{Ax_{n_k}\}$ is convergent.*

**Exercise 9.1.2.** Prove that a compact operator is bounded.

**Definition 9.1.3.** A bounded linear operator $A : X \to Y$ is said to have *finite rank* if Im $A$ is a finite dimensional subspace.

**Remark 9.1.4.** Note that we reserve the term *finite rank operator* to bounded operators. Unbounded operators which have a finite dimensional image are usually not referred to as finite rank operators; this is just a manner of terminological convenience.

**Example 9.1.5.** A finite rank operator $A$ is compact. Indeed, if $A$ is bounded, then $A(X_1)$ is a bounded subspace of the finite dimensional space Im $A$, hence by the Heine-Borel theorem (Theorem A.3.28) it is precompact.

In particular, operators on finite dimensional spaces are compact. It might seem like this would imply that the theory of compact operators is interesting only in the infinite dimensional setting. However, from the analysis of compact operators one may obtain elegant proofs for theorems pertaining to operators on finite dimensional spaces (see, for example, the Spectral Theorem for normal compact operators on a Hilbert space, Theorem 10.3.7).

**Example 9.1.6.** Let $k \in C([0,1]^2)$, and define an operator $K : C([0,1]) \to C([0,1])$ by

$$Kf(x) = \int_0^1 k(x,t)f(t)dt.$$

Then $K$ is a bounded operator with $\|K\| \le \|k\|_\infty$, as one may check directly. The image of the closed unit ball $C([0,1])_1$ under $K$ is bounded, by the boundedness of $K$. We will show that $K[C([0,1])_1]$ is *equicontinuous* — the Arzelà-Ascoli theorem then implies that $K[C([0,1])_1]$ is precompact (see Definition A.4.4 and Theorem A.4.6).

To establish equicontinuity, let $\epsilon > 0$ be given. Choose $\delta$ such that $|x-y| < \delta$ implies that $|k(x,t)-k(y,t)| < \epsilon$ for all $t$ (this is possible since $k$ is uniformly continuous on $[0,1]^2$; see Proposition A.3.29). For any $f \in C([0,1])_1$, we have

$$|Af(x) - Af(y)| \le \int_0^1 |k(x,t) - k(y,t)||f(t)|dt < \epsilon\|f\|_\infty \le \epsilon.$$

This shows that $A(C([0,1])_1)$ is equicontinuous, and we are done.

**Theorem 9.1.7.** *Let $X, Y$ and $Z$ be Banach spaces, and let $A, B \in B(X,Y)$, and $C \in B(Y,Z)$.*

1. *If $A$ is compact, then so is any scalar multiple of $A$.*

2. *If $A$ and $B$ are compact, then $A + B$ is compact.*

3. *If $A$ or $C$ are compact, then $CA$ is compact.*

4. *If $A_n$ is a sequence of compacts and $A_n \to A$, then $A$ is compact.*

*Proof.* Let $\{x_n\}$ be a sequence in $X_1$.

For (1) and (2), choose a subsequence $\{y_k\} = \{x_{n_k}\}$ such that $Ay_k$ converges. The sequence $\{y_k\}$ is still bounded, so choose a sub-subsequence such that $By_{k_l}$ converges. Then for any scalar $\lambda$,

$$Ay_{k_l} + \lambda By_{k_l}$$

converges.

If $A \in B(X,Y)$, then $\{Ax_n\}$ is bounded, so $\{CAx_n\}$ has a convergent subsequence if $C \in K(Y,Z)$. If, on the other hand $A \in K(X,Y)$ and $C \in B(Y,Z)$, then $\{Ax_n\}$ has a convergent subsequence, which remains convergent after applying $C$.

Finally, let $A_n$ be compacts converging to $A$. Given a sequence $\{x_n\}$ in $X_1$ we will construct a subsequence $\{y_n\}$ of $\{x_n\}$ such that $\{Ay_n\}$ is convergent. The construction of $\{y_n\}$ is done using an argument called a "diagonalization argument", which goes as follows.

Let $\{y_{1k}\}_k$ be a subsequence of $\{x_n\}$ such that $\{A_1y_{1k}\}$ is norm convergent. Supposing that we have defined sequences $\{y_{jk}\}_k$ for $j = 1, \ldots n-1$ such that $\{A_iy_{jk}\}_k$ converges for $i \le j$, we define inductively $\{y_{nk}\}_k$ to be a subsequence of $\{y_{(n-1)k}\}_k$ so that $\{A_n y_{nk}\}_k$ converges. We now define a sequence by $y_n = y_{nn}$ for all $n$. Then $\{Ay_n\}$ is Cauchy, hence convergent.

Indeed, fix $\epsilon > 0$, and let $N$ be such that $\|A_N - A\| < \epsilon$. We estimate

$$\|Ay_m - Ay_n\| \leq \|Ay_m - A_N y_m\| + \|A_N y_m - A_N y_n\| + \|A_N y_n - Ay_n\|$$
$$< \epsilon + \|A_N y_m - A_N y_n\| + \epsilon,$$

and the middle term tends to zero as $m, n \to \infty$. This is because $\{A_N y_n\}_n$ is a Cauchy sequence, for $\{y_n\}_n$ is eventually a subsequence of $\{y_{Nk}\}_k$, which was chosen so that $\{A_N y_{Nk}\}_k$ is convergent. That completes the proof. $\square$

**Exercise 9.1.8.** Find an alternative proof of the fact that the limit of compact operators is compact, using the characterization *a metric space is compact if and only if it is complete and totally bounded* (see Theorem A.3.27).

**Example 9.1.9.** Given $k \in L^2([0,1]^2)$, one can define an operator $K : L^2[0,1] \to L^2[0,1]$ by

$$Kf(x) = \int_0^1 k(x,t)f(t)dt \ , \ f \in L^2[0,1]. \tag{9.1}$$

The standard way to make sense of the above definition is to use measure theory. However, there is a way to understand such integral operators without knowing the measure theoretic definition of $L^2[0,1]$. To see how to do this, assume first that $k$ is continuous. In this case we can define the function $Kf$ to be given at the point $x$ by (9.1) whenever $f$ is a continuous function. Applying the Cauchy-Schwarz inequality for integrals of continuous functions, we see that $\|Kf\|_2 \leq \|k\|_2 \|f\|_2$, therefore (9.1) defines a bounded operator on $(C([0,1]), \|\cdot\|_2)$, and that already determines what the operator $K$ does on all of $L^2[0,1]$. Thus, when $k$ is continuous, we can define $K$ on $L^2[0,1]$ by first defining it on $C([0,1])$ according to (9.1), and the extending it to $L^2[0,1]$ using Proposition 5.1.6.

When $k$ is not continuous, one can find a sequence of continuous functions $k_n$ tending to $k$ in $L^2([0,1]^2)$. Then the corresponding operators $K_n$ form a Cauchy sequence, that converges in norm to the "integral" operator $K$, which is compact, thanks to Theorem 9.1.7 (see Exercise 10.5.3 for details).

## 9.2 The spectrum of a compact operator

Throughout this section, $X$ denotes a Banach space.

**Lemma 9.2.1.** *Let $Y$ be a closed subspace of $X$. Then if $Y \neq X$, there exists a unit vector $x \in X$ such that*

$$\inf\{\|x - y\| : y \in Y\} \geq 1/2.$$

*Proof.* Let $x_0 \in X \setminus Y$. Since $Y$ is closed, the distance $d$ between $x_0$ and $Y$ is positive:

$$d = \inf\{\|x_0 - y\| : y \in Y\} > 0. \tag{9.2}$$

Choose $y_0 \in Y$ such that $\|x_0 - y_0\| < 2d$. Define $x = \|x_0 - y_0\|^{-1}(x_0 - y_0)$. Then, using (9.2), we find that for all $y$

$$\|x - y\| = \|x_0 - y_0\|^{-1}\|x_0 - y_0 - y'\| > (2d)^{-1}d = 1/2,$$

where $y' = \|x_0 - y_0\|y$. That completes the proof. $\qquad\square$

**Exercise 9.2.2.** Prove that the inf in (9.2) is positive, and actually obtained when $Y$ is finite dimensional.

**Theorem 9.2.3.** *$X_1$ is compact if and only if* $\dim X < \infty$.

*Proof.* The fact that the unit ball is compact when the dimension of the space is finite is essentially the Heine-Borel theorem (Theorem A.3.28). We will show that if $X$ is infinite dimensional, then there is a sequence of points $x_n \in X_1$ such that $\|x_n - x_m\| \geq 1/2$ for all $m \neq n$ — clearly this implies that $X_1$ is not compact.

We construct the sequence inductively, starting from an arbitrary unit vector $x_1$. Having chosen $x_1, \ldots x_n$, we define $Y = \text{span}\{x_1, \ldots, x_n\}$, and apply the above lemma to find $x_{n+1}$, as required. $\qquad\square$

**Corollary 9.2.4.** *The identity operator $I_X$ on a Banach space $X$ is compact if and only if* $\dim X < \infty$.

*Proof.* This is immediate from the above theorem, since $I_X(X_1) = X_1$. $\qquad\square$

**Theorem 9.2.5.** *Let $A$ be a compact operator between Banach spaces $X$ and $Y$. If either $X$ or $Y$ is infinite dimensional, then $A$ is not invertible. In particular, if $A \in K(X)$ and $\dim X = \infty$, then*

$$0 \in \sigma(A).$$

*Proof.* If $A$ is invertible and compact, then $A^{-1}A = I_X$ is compact (by Theorem 9.1.7), thus $\dim X$ must be finite by the above corollary. $\qquad\square$

We see that the spectrum of a compact operator (on an infinite dimensional space) always contains 0. Our next goal is to understand the nonzero points in the spectrum. Recall the notion of an operator being bounded below (Definition 8.3.2).

**Lemma 9.2.6.** *Let $A \in K(X)$, let $Y \subseteq X$ be a closed subspace, and assume that $\ker(I - A) \cap Y = \{0\}$. Then, the restriction $(I - A)|_Y$ of $I - A$ to $Y$ is bounded below, and, consequently, $(I - A)Y$ is closed.*

*Proof.* We will show that if $(I - A)|_Y$ is not bounded below, then $\ker(I - A) \cap Y \neq \{0\}$. If $(I - A)|_Y$ is not bounded below, then there is a sequence $\{y_n\}$ of unit vectors in $Y$ such that $\lim(I - A)y_n = 0$. Since $A$ is compact, there is a subsequence $\{y_{n_k}\}$ such that $Ay_{n_k}$ is convergent. But then

$$y_{n_k} = Ay_{n_k} + (I - A)y_{n_k} \tag{9.3}$$

is also convergent to some $y \in Y$, and note that $\|y\| = 1$. Taking the limit as $k \to \infty$ we find that $(I - A)y = \lim_k (I - A)y_{n_k} = 0$, so $\ker(I - A) \cap Y \neq \{0\}$.

The closedness of $(I - A)Y$ now follows readily from boundedness below (see Exercise 8.3.3). $\qquad\square$

**Lemma 9.2.7.** *Let $A \in K(X)$. If $\ker(I - A) = \{0\}$, then $\text{Im}(I - A) = X$.*

**Remark 9.2.8.** The converse also holds, but we shall not prove it in this generality. In Corollary 11.1.6, the converse will be obtained in the setting of Hilbert spaces.

*Proof.* Let us define $Y_n = (I - A)^n X$. Then we have $X = Y_0 \supseteq Y_1 \supseteq Y_2 \supseteq \ldots$, and by the previous lemma, each $Y_n$ is a closed subspace of $Y_{n-1}$. We need to show that $Y_1 = X$.

Assume for contradiction that $Y_1 \subsetneq X$. This implies that $Y_n \subsetneq Y_{n-1}$ for all $n \geq 1$. Indeed, if $x \in X \backslash \text{Im}(I-A)$, then $(I-A)x \in (I-A)X \backslash (I-A)^2 X$ — this is because if $(I-A)x = (I-A)^2 y$, we would have that $x - y + Ay \in \ker(I-A)$, so by injectivity $x = (I - A)y \in \text{Im}(I - A)$; this shows that $Y_1 \subsetneq X$ implies $Y_2 \subsetneq Y_1$. Inductively, if $Y_1 \subsetneq X$ then $Y_n \subsetneq Y_{n-1}$ holds for all $n \geq 1$.

Thus we obtain a strictly decreasing sequence of subspaces $Y_0 \supsetneq Y_1 \supsetneq Y_2 \supsetneq \ldots$. By Lemma 9.2.1 there is a sequence of unit vectors $\{y_n\}$, with $y_n \in Y_n$, such that $\|y_m - y\| \geq 1/2$ for all $m < n$ and all $y \in Y_n$. But for all $m < n$,

$$Ay_m - Ay_n = y_m - ((I - A)y_m + Ay_n)$$

and the element $y = (I - A)y_m + Ay_n$ is contained in $Y_{m+1}$ (why?). Therefore $\|Ay_m - Ay_n\| \geq 1/2$, and this shows that $\{Ay_n\}$ has no Cauchy subsequence. This contradicts the assumption that $A$ is compact, so the proof is complete. $\qquad\square$

**Theorem 9.2.9** (The Fredholm alternative). *Let $A \in K(X)$ and fix a complex number $\lambda \neq 0$. Then $\dim \ker(\lambda - A)$ is finite dimensional, and exactly one of the following holds:*

*1. $\ker(\lambda - A) \neq \{0\}$,*

*2. $\lambda - A$ is invertible.*

*Thus, every nonzero point in $\sigma(A)$ is an eigenvalue of finite multiplicity. In particular, if $X$ is an infinite dimensional Banach space, then*

$$\sigma(A) = \sigma_p(A) \cup \{0\}.$$

*Proof.* If $\lambda \neq 0$, and if we put $N = \ker(\lambda - A)$, then the restriction operator $A\big|_N$ is on one hand compact (why?) and on the other hand it is equal to the invertible operator $\lambda I_N$. It follows from Lemma 9.2.5 that $\dim N < \infty$.

Next, for $\lambda \neq 0$, we have that $\ker(\lambda - A) = \ker(I - \lambda^{-1}A)$, and $\lambda - A$ is invertible if and only if $I - \lambda^{-1}A$ is invertible. Moreover, $A$ is compact if and only if $\lambda^{-1}A$ is compact. Thus, to prove the dichotomy between (1) and (2), it suffices to show that if $\ker(I - A) = \{0\}$, then $I - A$ is invertible.

So suppose that $\ker(I - A) = \{0\}$. By Lemma 9.2.7, $I - A$ is surjective, so it has linear inverse $(I - A)^{-1}$. By Lemma 9.2.6, $I - A$ is bounded below, so $(I - A)^{-1}$ is bounded, and therefore $I - A$ is invertible. $\qquad\square$

Thus, an operator of the form "identity plus compact" is invertible if and only if it is injective. The same is true for an operator of the form "invertible plus compact".

**Corollary 9.2.10.** *Let $X$ be a Banach space, let $T \in GL(X)$ and let $A \in K(X)$. If $T + A$ is injective, then it is invertible.*

**Exercise 9.2.11.** Prove the corollary.

Another important piece of information regarding the spectral theory of compact operators is the following theorem.

**Theorem 9.2.12.** *Let $A \in K(X)$. The nonzero points in $\sigma(A)$ form a countable sequence converging to 0.*

*Proof.* Fix $\epsilon > 0$. We will show that the existence of a sequence $\{\lambda_n\}_{n=1}^{\infty}$ of distinct elements in $\sigma(A)$, such that $|\lambda_n| > \epsilon$ for all $n$, leads to a contradiction. Indeed, the existence of such a sequence implies the existence of a corresponding sequence of unit eigenvectors $\{x_n\} \subseteq X$. We know from elementary linear algebra (and from Exercise 8.5.15) that the $x_n$s must be linearly independent. Thus, we may form the spaces

$$Y_n = \mathrm{span}\{x_1, \ldots, x_n\}.$$

Using Lemma 9.2.1 we form a sequence $\{y_n\}$ of unit vectors such that $\|y_n - y\| \geq 1/2$ for all $y \in Y_{n-1}$. We leave it to the reader to argue (similarly to the proof of Lemma 9.2.7) that this implies that $\{Ay_n\}$ contains no Cauchy subsequence, and this contradicts compactness. $\qquad\square$

**Exercise 9.2.13.** Complete the details of the above proof.

## 9.3 Additional exercises

**Exercise 9.3.1.** Let $K$ be the integral operator defined in Example 9.1.6. Calculate $\|K\|$ in terms of $k$.

**Exercise 9.3.2.** Let $A \in K(X,Y)$. Prove that if $Z$ is a subspace of $X$, then $A\big|_Z$ is compact.

**Exercise 9.3.3.** Let $X$ be an infinite dimensional Banach space, and let $f(z) = \sum_{n=0}^{\infty} a_n z^n$ be an analytic function in $D_r(0)$. Let $A$ be a compact operator with $\|A\| < r$, and define $f(A)$ as in Exercise 8.5.11. Prove that $f(A)$ is compact if and only if $a_0 = 0$.

**Exercise 9.3.4** (The Volterra operator). Define *the Volterra operator* $V$ : $C([0,1]) \to C([0,1])$ by

$$Vf(x) = \int_0^x f(t)dt.$$

Prove that $V$ is compact, and calculate the spectrum of $V$. (Note that this is not exactly the operator from Example 9.1.6.)

**Exercise 9.3.5** (Volterra type operator). A *Volterra type operator* $T$ : $C([0,1]) \to C([0,1])$ is an operator of the form

$$Tf(x) = \int_0^x k(x,t)f(t)dt,$$

where $k$ is a continuous function on $[0,1]^2$ (equivalently, one may take a continuous function on $\{(x,t) : 0 \le t \le x \le 1\}$). This is not exactly the type of operator treated in Example 9.1.6. The Volterra operator from Exercise 9.3.4 is a Volterra type operator, with $k \equiv 1$. Prove that every Volterra type operator is compact, and has the same spectrum as the Volterra operator.

**Exercise 9.3.6.** Let $C([0,1])$ equipped with the norm $\|f\|_\infty = \sup_{t\in[0,1]} |f(t)|$, and let $C^1([0,1])$ be the space of all continuously differentiable functions on $[0,1]$, equipped with the norm

$$\|f\|_{C^1} = \|f\|_\infty + \|f'\|_\infty.$$

This is a Banach space (recall Exercise 7.5.3). Determine which of the following operators is compact:

1. $A : C([0,1]) \to C^1([0,1])$ given by

$$Af(x) = \int_0^x f(t)dt.$$

2. $B : C^1([0,1]) \to C([0,1])$ given by

$$Bf(x) = \int_0^x f(t)dt.$$

3. $C : C^1([0,1]) \to C^1([0,1])$ given by

$$Cf(x) = \int_0^x f(t)dt.$$

4. $D : C^1([0,1]) \to C([0,1])$ given by

$$Df = \frac{df}{dx}.$$

5. $E : C^1([0,1]) \to C([0,1])$ given by

$$Ef = f.$$

**Exercise 9.3.7.** Let $p \in [1,\infty]$. Let $(w_n)_n$ be a convergent sequence, and define an operator $A : \ell^p \to \ell^p$ by

$$A(x_n) = (w_n x_n).$$

1. Determine precisely when $A$ is compact (make sure that you take care of all possible values of $p$).

2. Find the eigenvalues of $A$, and use the Fredholm alternative to compute the spectrum of $A$.

3. What can you say about the spectrum of $A$ if $(w_n)$ is a bounded sequence which is not assumed convergent?

**Exercise 9.3.8.** Let $A = (a_{ij})_{i,j=0}^{\infty}$ be an infinite matrix. For each of the following conditions, determine whether it is necessary, sufficient, or neither, so that matrix multiplication by $A$ defines a compact operator on $\ell^1$:

1. $\lim_{j\to\infty} a_{ij} = 0$ for all $i \in \mathbb{N}$.

2. $\lim_{i\to\infty} a_{ij} = 0$ for all $j \in \mathbb{N}$.

3. $\lim_{j\to\infty} \sum_{i=0}^{\infty} |a_{ij}| = 0$.

4. $\lim_{i\to\infty} \sum_{j=0}^{\infty} |a_{ij}| = 0$.

5. $\sum_{i,j=0}^{\infty} |a_{ij}| < \infty$.

**Exercise 9.3.9.** Let $\alpha : [0,1] \to [0,1]$ be a continuously differentiable map, and define a composition operator $T : f \mapsto f \circ \alpha$ on $L^p[0,1]$ ($p \in [1,\infty)$) similarly to as in Example 7.2.10. Prove that $T$ is bounded, and determine for which $\alpha$ is $T$ a compact operator.

**Exercise 9.3.10.** Find all continuous functions $g$ on $[0,1]$, such that the multiplication operator $M_g$ (as defined in Example 5.1.7) is compact on $L^2[0,1]$.

**Exercise 9.3.11.** Let $S$ denote the right shift operator on $\ell^2$. Does there exist a compact operator $A \in B(\ell^2)$ such that $S + A$ is invertible?

# Chapter 10

## Compact operators on Hilbert space

In the previous chapter, we introduced and studied compact operators on Banach spaces. In this chapter, we will restrict attention to compact operators on Hilbert spaces, where, due to the unique geometry of Hilbert spaces, we will be able to say much more. The culmination of this chapter is the so-called *spectral theorem* for compact normal operators on a Hilbert space. The spectral theorem is the infinite dimensional version of the familiar fact that every normal matrix is unitarily diagonalizable.

## 10.1   Finite rank operators on Hilbert space

Let $H$ and $K$ be two Hilbert spaces. If $h \in H$ and $k \in K$, then we define the **rank one operator** $k \otimes h^*$ to be the operator given by

$$k \otimes h^*(g) = \langle g, h \rangle k.$$

**Exercise 10.1.1.** Prove the following.

1. $k \otimes h^*$ is a bounded operator and $\|k \otimes h^*\| = \|h\| \|k\|$.

2. $\dim(\mathrm{Im}[k \otimes h^*]) = 1$ (unless one of the vectors is zero).

3. $(k \otimes h^*)^* = h \otimes k^*$.

4. For every operator $A$ defined on $K$, $A(k \otimes h^*) = (Ak) \otimes h^*$.

5. For every operator $B$ defined on $H$, $(k \otimes h^*)B^* = k \otimes (Bh)^*$.

Now if $h_1, \ldots, h_n \in H$ and $k_1, \ldots, k_n \in K$, the operator $\sum_{i=1}^{n} k_i \otimes h_i^*$ is clearly a finite rank operator. Every finite rank operator arises this way.

**Theorem 10.1.2.** *Let $F \in B(H, K)$ be a finite rank operator (i.e., $\dim \mathrm{Im}\, F < \infty$). Then there exists $h_1, \ldots, h_n \in H$ and $k_1, \ldots, k_n \in K$ such that*

$$F = \sum_{i=1}^{n} k_i \otimes h_i^*.$$

*For the vectors $k_1, \ldots, k_n$ one can take an orthonormal basis for $\mathrm{Im}\, F$.*

*Proof.* Let $k_1, \ldots, k_n$ be an orthonormal basis for Im $F$. Then for every $h \in H$,

$$Fh = \sum_{i=1}^{n} \langle Fh, k_i \rangle k_i.$$

Putting $h_i = F^* k_i$ for $i = 1, \ldots, n$ completes the proof. $\qquad\square$

From the theorem and Exercise 10.1.1, one can easily prove the following.

**Exercise 10.1.3.** The set of finite rank operators on $H$ is a subspace of $B(H)$, and if $F$ is a finite rank operator and $A \in B(H)$ then $F^*$, $AF$ and $FA$ are all finite rank operators.

**Example 10.1.4.** Let $K$ be the operator on $L^2[0, 2\pi]$ given by

$$Kf(x) = \int_0^{2\pi} \cos(x + t) f(t) dt.$$

According to Example 9.1.9 this integral operator is compact, because $\cos(x + t) \in L^2\left([0, 2\pi]^2\right)$. It turns out that it is a finite rank operator. Indeed, using $\cos(x + t) = \cos(x)\cos(t) - \sin(x)\sin(t)$, we have

$$Kf(x) = \int_0^{2\pi} \cos(x)\cos(t) f(t) dt - \int_0^{2\pi} \sin(x)\sin(t) f(t) dt$$

so $K$ is the finite rank operator $k_1 \otimes h_1^* + k_2 \otimes h_2^*$, where $k_1 = h_1 = \cos(x)$, $k_2 = -h_2 = \sin(x)$.

**Example 10.1.5.** The general form of a finite rank operator on $L^2[0, 2\pi]$ is

$$Kf(x) = \sum_{i=1}^{n} \left( \int_0^{2\pi} \overline{g_i(t)} f(t) dt \right) h_i(x).$$

This is the integral operator $Kf(x) = \int_0^{2\pi} k(x, t) f(t) dt$ with kernel $k(x, t) = \sum_{i=1}^{n} \overline{g_i(t)} h_i(x)$.

---

## 10.2    The spectral theorem for compact self-adjoint operators

In the previous chapter, we saw that every nonzero element of the spectrum of a compact operator on a Banach space is an eigenvalue. On the other hand, there do exist compact operators that have no eigenvalues (as Exercise 9.3.4 shows) and such operators cannot be understood well in terms of their spectrum. In this section and the next one, we turn to *self-adjoint* compact operators on a Hilbert space. We shall see that such operators can be reconstructed from their spectral data.

**Definition 10.2.1.** An operator $A \in B(H)$ is said to be **diagonal** if there is an orthonormal basis $\{e_i\}_{i \in I}$ for $H$ and a family of scalars $\{\lambda_i\}_{i \in I}$ such that for all $h \in H$,

$$Ah = A \left( \sum_{i \in I} \langle h, e_i \rangle e_i \right) = \sum_{i \in I} \lambda_i \langle h, e_i \rangle e_i. \tag{10.1}$$

It is easy to see that a scalar $\lambda$ is an eigenvalue of a diagonal operator $A$ as in the definition if and only if it appears as one of the $\lambda_i$s.

**Definition 10.2.2.** Let $\{\lambda_i\}_{i \in I}$ be a family of scalars. We say that $\{\lambda_i\}_{i \in I}$ **converges to** 0 if for every $\epsilon > 0$, there exists a finite subset $F \subseteq I$ such that $|\lambda_i| < \epsilon$ for all $i \notin F$.

Note carefully that by the above definition a finite family of scalars $\{\lambda_i\}_{i=1}^N$ converges to 0.

**Exercise 10.2.3.** Show that the following conditions are equivalent:

1. $\{\lambda_i\}_{i \in I}$ converges to 0,

2. One of the following holds: **(a)** the set $J = \{i \in I : \lambda_i \neq 0\}$ is finite; or **(b)** $J = \{i \in I : \lambda_i \neq 0\}$ is countably infinite, and $\{\lambda_j\}_{j \in J}$ is a sequence convergent to 0 in the ordinary sense (it does not matter in which order it is taken).

**Exercise 10.2.4.** Prove that a diagonal operator $A$ given by (10.1) is compact if and only if $\{\lambda_i\}_{i \in I}$ converges to 0, and that this happens if and only if the series $\sum_{i \in I} \lambda_i e_i \otimes e_i^*$ converges in the operator norm to $A$.

(To say that $\sum_{i \in I} \lambda_i e_i \otimes e_i^*$ converges in the operator norm to $A$ means that for every $\epsilon > 0$, there exists a finite set $F_0 \subset I$, such that $\| \sum_{i \in F} \lambda_i e_i \otimes e_i^* - A \| < \epsilon$ for all finite $F \supseteq F_0$; equivalently, if $J = \{j_1, j_2, \ldots\}$ denotes the set of indices $i$ for which $\lambda_i \neq 0$, then $\lim_{N \to \infty} \sum_{n=1}^N \lambda_{j_n} e_{j_n} \otimes e_{j_n}^* = A$ in norm).

**Definition 10.2.5.** The **numerical radius** of an operator $A \in B(H)$ is the number
$$r(A) = \sup\{|\langle Ah, h \rangle| : \|h\| = 1\}.$$

**Proposition 10.2.6.** *For every* $A \in B(H)$,
$$r(A) \leq \|A\|.$$

*If $A$ is self-adjoint, then*
$$\|A\| = r(A).$$

*Proof.* The inequality is immediate from $\|A\| = \sup_{\|h\|=\|g\|=1} |\langle Ah, g \rangle|$ and Cauchy-Schwarz. If $A$ is self-adjoint, we have

$$4\operatorname{Re}\langle Ah, g \rangle = \langle A(h + g), h + g \rangle - \langle A(h - g), h - g \rangle,$$

which gives (using the parallelogram law)

$$4|\text{Re}\langle Ah, g\rangle| \leq r(A) \left( \|h + g\|^2 + \|h - g\|^2 \right)$$
$$= 2r(A) \left( \|h\|^2 + \|g\|^2 \right).$$

Since this is true for all $g, h \in H$, we can replace $g$ by $e^{i\theta}g$ for some $\theta \in \mathbb{R}$, and we reach

$$4|\langle Ah, g\rangle| \leq 2r(A) \left( \|h\|^2 + \|g\|^2 \right),$$

for all $g, h$. Taking the supremum over $h, g \in H_1$ gives $\|A\| \leq r(A)$.   □

**Lemma 10.2.7.** *Let $A$ be a self-adjoint operator on a Hilbert space $H$. Then, there exists a point $\lambda \in \sigma(A)$ for $A$ such that either $\lambda = \|A\|$ or $\lambda = -\|A\|$. In particular, the spectrum of a selfadjoint operator is not empty. If $A$ is also compact, then there is an eigenvalue $\lambda \in \sigma_p(A)$ for $A$ such that either $\lambda = \|A\|$ or $\lambda = -\|A\|$.*

*Proof.* Suppose that $A \neq 0$, otherwise the result is immediate. Let $\{h_n\}$ be unit vectors such that $\lim_n \langle Ah_n, h_n \rangle = \lambda \in \mathbb{R}$, where $|\lambda| = r(A) = \|A\|$. We first show that $\{h_n\}$ are approximately eigenvectors, in the sense that $Ah_n - \lambda h_n \to 0$. Indeed,

$$0 \leq \|Ah_n - \lambda h_n\|^2$$
$$= \|Ah_n\|^2 - 2\text{Re}\lambda\langle Ah_n, h_n \rangle + \lambda^2$$
$$\leq \lambda^2 - 2\text{Re}\lambda\langle Ah_n, h_n \rangle + \lambda^2 \to 0.$$

From this it follows that $A - \lambda$ is not bounded below, and therefore $\lambda \in \sigma(A)$.

Now assume that $A$ is compact, and let $\{h_{n_k}\}$ be a subsequence such that $\{Ah_{n_k}\}$ is a convergent sequence, say $Ah_{n_k} \to g$. Then $\lambda h_{n_k} - g \to 0$, thus $\|g\| = |\lambda| \neq 0$ and

$$Ag = \lim_k \lambda Ah_{n_k} = \lambda g.$$

□

**Definition 10.2.8.** Let $A \in B(H)$. A subspace $M \subseteq H$ is said to be ***invariant*** for $A$ if $AM \subseteq M$. A subspace $M \subseteq H$ is said to be ***reducing*** for $A$ if $M$ and $M^\perp$ are invariant for $A$; in this case we also say that $M$ ***reduces*** $A$.

**Exercise 10.2.9.** Let $A \in B(H)$, $M \subseteq H$ a closed subspace and let $P_M$ be the orthogonal projection onto $M$. Prove the following facts.

- $M$ is invariant for $A$ if and only if $AP_M = P_M AP_M$.

- $M$ is reducing for $A$ if and only if $AP_M = P_M A$.

- $M$ is reducing for $A$ if and only if $M$ is invariant for both $A$ and $A^*$.

- If $A$ is self-adjoint, then every invariant subspace $M$ for $A$ is reducing, and $A\big|_M$ (the restriction of $A$ to $M$) is also a self-adjoint operator.

- If $A$ is compact and $M$ is an invariant subspace, then $A\big|_M$ is a compact operator on $M$.

**Lemma 10.2.10.** *Let $A$ be a self-adjoint operator on a Hilbert space $H$. For every $\lambda \in \sigma_p(A)$, the space $\ker(\lambda - A)$ is reducing for $A$.*

*Proof.* Every eigenspace is invariant, hence reducing by the exercise above. $\square$

**Theorem 10.2.11** (The spectral theorem for compact self-adjoint operators). *Let $A$ be a compact, self-adjoint operator on a Hilbert space $H$. Then, there exists an orthonormal basis $\{e_i\}_{i\in I}$ for $H$ and a family $\{\lambda_i\}_{i\in I}$ of real numbers, convergent to $0$, such that*

$$Ah = \sum_{i\in I} \lambda_i \langle h, e_i \rangle e_i, \tag{10.2}$$

*for all $h \in H$.*

*Proof.* Let $\lambda_1$ be an eigenvalue such that $|\lambda_1| = \|A\|$, and let $e_1$ be a corresponding unit eigenvector (Lemma 10.2.7). Letting $M_1 = \operatorname{span}\{e_1\}$, we have that $M_1$ reduces $A$, and that $A\big|_{M_1}$ and $A\big|_{M_1^\perp}$ are compact self-adjoint operators. If $H$ is finite dimensional, the proof is completed by induction on $\dim H$; if $H$ is infinite, then we proceed as follows.

We continue to define the subspaces $M_n$ with orthonormal basis $\{e_1, \ldots, e_n\}$ recursively: if $M_n = \operatorname{span}\{e_1, \ldots, e_n\}$ we consider the compact self-adjoint operator $A' = A\big|_{M_n^\perp}$, and find an eigenvalue $\lambda_{n+1}$ such that $|\lambda_{n+1}| = \|A'\|$ with corresponding unit eigenvector $e_{n+1}$, and we define $M_{n+1} = \operatorname{span}\{e_1, \ldots, e_{n+1}\}$.

Note that $|\lambda_{n+1}| \leq |\lambda_n|$ for all $n$. We must have that $\lambda_n \to 0$. This follows from the spectral theory of compact operators on a Banach space (Theorem 9.2.12), but in this special situation we may give a simpler argument. Indeed, if $\lambda_n \not\to 0$, then there would be infinitely many values of $m, n$ such that

$$\|Ae_n - Ae_m\|^2 = \|\lambda_n e_n - \lambda_m e_m\|^2 = \lambda_n^2 + \lambda_m^2,$$

remains bounded away from $0$, contradicting compactness of $A$.

For every $n$ we have that $h - \sum_{i=1}^n \langle h, e_i \rangle e_i = P_{M_n^\perp} h$, so

$$\left\| Ah - \sum_{i=1}^n \lambda_i \langle h, e_i \rangle e_i \right\| = \left\| A\big|_{M_n^\perp} P_{M_n^\perp} h \right\|$$

$$\leq \left\| A\big|_{M_n^\perp} \right\| \|h\|$$

$$= |\lambda_{n+1}| \|h\| \to 0.$$

This shows that

$$Ah = \lim_{n\to\infty} \sum_{i=1}^n \lambda_i \langle h, e_i \rangle e_i.$$

To obtain (10.2) we complete $\{e_i\}_{i=1}^{\infty}$ to an orthonormal basis $\{e_i\}_{i\in I}$ for $H$, and define $\lambda_i = 0$ for $i \in I \setminus \{1, 2, \ldots\}$ (note that perhaps $\lambda_i = 0$ already for some $i \in \{1, 2, \ldots\}$, but this causes no problems). $\quad\square$

Recall that the nonzero points of the spectrum of a compact operator consist of eigenvalues. If we define $N_\lambda = \ker(\lambda - A)$, we can reformulate the above theorem by saying that for every compact self-adjoint operator, $H$ decomposes as the direct sum

$$H = \bigoplus_{\lambda \in \sigma(A)} N_\lambda = \bigoplus_{\lambda \in \sigma_p(A)} N_\lambda, \tag{10.3}$$

where every $N_\lambda$ is reducing for $A$, and $A$ acts on $N_\lambda$ as the scalar operator $\lambda I_{N_\lambda}$ (note that if 0 is in $\sigma(A)$ but not in $\sigma_p(A)$, then $N_0$ is the zero subspace, which contributes nothing to the sum). We may therefore say that $A$ can be represented as a **direct sum**, or as a **block diagonal operator** corresponding to the direct sum (10.3)

$$A = \oplus_{\lambda \in \sigma_p(A)} \lambda P_{N_\lambda},$$

(recall Exercise 5.7.15). We thus obtain the following reformulation of the spectral theorem.

**Theorem 10.2.12.** *Let $A$ be a compact, self-adjoint operator on a Hilbert space $H$. Then*

$$Ah = \sum_{\lambda \in \sigma_p(A)} \lambda P_{N_\lambda} h \tag{10.4}$$

*for all $h \in H$, where $N_\lambda = \ker(\lambda - A)$ and $P_{N_\lambda}$ is the orthogonal projection on $N_\lambda$.*

## 10.3 The spectral theorem for compact normal operators

**Theorem 10.3.1.** *Let $A \in B(H, K)$. Then $A$ is compact if and only if $A^*$ is compact.*

*Proof.* By the symmetry of the adjoint operation, we need only prove one direction. Assume that $A$ is compact, and let $\{k_n\}$ be a bounded sequence in $K$. Since $A$ is compact and $A^*$ is bounded, there is a subsequence $\{k_{n_j}\}_j$ such that $AA^* k_{n_j}$ is convergent, hence

$$\|A^* k_{n_i} - A^* k_{n_j}\|^2 = \langle AA^*(k_{n_i} - k_{n_j}), k_{n_i} - k_{n_j} \rangle$$
$$\leq \|AA^*(k_{n_i} - k_{n_j})\| \|k_{n_i} - k_{n_j}\| \to 0.$$

Thus, $\{A^* k_{n_j}\}_j$ is a convergent subsequence of $\{A^* k_n\}$. This shows that $A^*$ is compact. $\quad\square$

**Lemma 10.3.2.** *Every $A = B(H)$ can be written in a unique way as the sum*

$$A = A_1 + iA_2,$$

*where $A_1, A_2$ are self-adjoint. Moreover, $A$ is compact if and only if both $A_1$ and $A_2$ are.*

*Proof.* We put $A_1 = \frac{A+A^*}{2}$ and $A_2 = \frac{A-A^*}{2i}$; this shows that there is such a decomposition and also shows (thanks to Theorem 10.3.1) that the $A_i$s are compact if and only if $A$ is. If $A = B_1 + iB_2$ is another such representation, then $\frac{A+A^*}{2} = B_1$, so the representation is unique. □

**Definition 10.3.3.** Two operators $A, B \in B(H)$ are said to **commute** if

$$AB = BA.$$

**Lemma 10.3.4.** *Let $A \in B(H)$, and let $A = A_1 + iA_2$ where $A_1, A_2$ are self-adjoint. Then $A$ is normal if and only if $A_1$ and $A_2$ commute.*

*Proof.* Let us introduce the notation $[A_1, A_2] = A_1 A_2 - A_2 A_1$ (the object $[A_1, A_2]$ is called the **commutator** of $A_1$ and $A_2$). A computation shows that

$$AA^* = (A_1 + iA_2)(A_1 - iA_2) = A_1^2 + A_2^2 - i[A_1, A_2],$$

and

$$A^*A = (A_1 - iA_2)(A_1 + iA_2) = A_1^2 + A_2^2 + i[A_1, A_2].$$

Thus $A$ is normal if and only if $[A_1, A_2] = -[A_1, A_2]$, that is, if and only if $[A_1, A_2] = 0$, as required. □

**Lemma 10.3.5.** *Let $A$ and $B$ be a pair of commuting self-adjoint operators on a Hilbert space $H$. Then every eigenspace of $A$ is a reducing subspace for $B$.*

*Proof.* Let $\lambda \in \sigma_p(A)$ and let $g \in N_\lambda$ be a corresponding eigenvector. Then $ABg = BAg = \lambda Bg$, thus $Bg \in N_\lambda$. We conclude that $N_\lambda$ is invariant under $B$. Since $B$ is self-adjoint, $N_\lambda$ is reducing for $B$. □

**Lemma 10.3.6.** *Let $A$ and $B$ be a pair of commuting compact self-adjoint operators on a Hilbert space $H$. Then $A$ and $B$ are simultaneously diagonalizable, in the sense that there exists an orthonormal basis $\{e_i\}_{i \in I}$ for $H$, and two families of real numbers $\{\lambda_i\}_{i \in I}$ and $\{\mu_i\}_{i \in I}$, such that*

$$Ah = \sum_{i \in I} \lambda_i \langle h, e_i \rangle e_i \quad and \quad Bh = \sum_{i \in I} \mu_i \langle h, e_i \rangle e_i,$$

*for all $h \in H$.*

*Proof.* By Theorem 10.2.12, $A$ is given by

$$Ah = \sum_{\lambda \in \sigma_p(A)} \lambda P_{N_\lambda} h$$

for all $h \in H$, where $N_\lambda = \ker(\lambda - A)$.

Let $\lambda \in \sigma_p(A)$. By the previous lemma, $N_\lambda$ is reducing for $B$, and we obtain that $B\big|_{N_\lambda}$ is a compact, self-adjoint operator on $N_\lambda$. Now, on $N_\lambda$ the operator $A$ acts as a scalar, thus it is diagonal with respect to any basis. Using the spectral theorem for compact self-adjoint operators (applied to $B\big|_{N_\lambda}$), we may find an orthonormal basis $\{e_i^\lambda\}_{i \in I_\lambda}$ for every $N_\lambda$ with respect to which both $A\big|_{N_\lambda}$ and $B\big|_{N_\lambda}$ are diagonal. Now $\{e_i^\lambda : i \in I_\lambda, \lambda \in \sigma_p(A)\}$ is an orthonormal basis with respect to which both $A$ and $B$ are diagonal. $\qquad\square$

**Theorem 10.3.7** (The spectral theorem for compact normal operators). *Let $A$ be a compact normal operator on $H$. Then, there exists an orthonormal basis $\{e_i\}_{i \in I}$ for $H$, and a family $\{\alpha_i\}_{i \in I}$ of complex scalars convergent to $0$, such that*

$$Ah = \sum_{i \in I} \alpha_i \langle h, e_i \rangle e_i,$$

*for all $h \in H$.*

*Proof.* Write $A = A_1 + iA_2$ with $A_1, A_2$ compact and self-adjoint as in Lemma 10.3.2. By Lemma 10.3.4 $A_1$ and $A_2$ commute, so the above lemma shows that they are simultaneously diagonalizable, say

$$A_1 h = \sum_{i \in I} \lambda_i \langle h, e_i \rangle e_i \quad \text{and} \quad A_2 h = \sum_{i \in I} \mu_i \langle h, e_i \rangle e_i,$$

for all $h \in H$. Then we have that

$$Ah = \sum_{i \in I} (\lambda_i + i\mu_i) \langle h, e_i \rangle e_i$$

for all $h \in H$, and moreover $\alpha_i = \lambda_i + i\mu_i$ is convergent to $0$. $\qquad\square$

## 10.4 The functional calculus for compact normal operators

In Exercise 8.5.11 we saw how to "evaluate a function at an operator", that is, how to make sense of $g(A)$ when $g$ is an analytic function on a disc $D_r(0)$, with $r > \|A\|$. The term *functional calculus* refers to the procedure of evaluating a function $g$ (from within a specified class of functions) at an operator $A$ (from within a specified class of operators), to form a new operator $g(A)$. This notion also goes under the names *function calculus, operational calculus*, or *function of an operator*. The purpose of this section is to introduce the functional calculus for compact normal operators.

Let $A$ be a compact normal operator on a Hilbert space $H$. For simplicity, we assume that $\dim H = \aleph_0$. By Theorem 10.3.7, there is an orthonormal basis $\{f_n\}_{n \in \mathbb{N}}$ consisting of eigenvectors with corresponding eigenvalues $\{c_n\}_{n \in \mathbb{N}}$, and $A$ is given by

$$Ah = A \sum_{n \in \mathbb{N}} \langle h, f_n \rangle f_n = \sum_{n \in \mathbb{N}} c_n \langle h, f_n \rangle f_n.$$

By Section 5.6, $A$ is unitarily equivalent to its matrix representation with respect to $\{f_n\}_{n \in \mathbb{N}}$, which is diagonal:

$$[A] = \begin{pmatrix} c_0 & & & \\ & c_1 & & \\ & & c_2 & \\ & & & \ddots \end{pmatrix}.$$

If $U : H \to \ell^2$ is the unitary that maps $f_n$ to the $n$th standard basis vector $e_n \in \ell^2$, then

$$[A] = UAU^*.$$

If $g(z) = \sum_{n=0}^{N} a_n z^n$ is a polynomial, then we know how to define $g(A)$, and $[g(A)] = [\sum a_n A^n] = \sum a_n [A]^n$. Therefore,

$$[g(A)] = \sum_{n=0}^{N} a_n \begin{pmatrix} c_0 & & & \\ & c_1 & & \\ & & c_2 & \\ & & & \ddots \end{pmatrix}^n = \begin{pmatrix} g(c_0) & & & \\ & g(c_1) & & \\ & & g(c_2) & \\ & & & \ddots \end{pmatrix}.$$

This extends also to analytic functions defined on a disc $D_r(0)$ with $r > \|A\|$.

**Definition 10.4.1.** Let $A$ be a compact normal operator as above, and let $g : \sigma_p(A) \to \mathbb{C}$ be a bounded function defined on $\sigma_p(A)$. We define $g(A)$ to be the bounded operator on $H$ given by

$$g(A)h = \sum_{n \in \mathbb{N}} g(c_n) \langle h, f_n \rangle f_n,$$

that is, the operator with matrix representation

$$[g(A)] = \begin{pmatrix} g(c_0) & & & \\ & g(c_1) & & \\ & & g(c_2) & \\ & & & \ddots \end{pmatrix},$$

with respect to the orthonormal basis $\{f_n\}_{n \in \mathbb{N}}$.

It is not hard to see that $g(A)$ is a well-defined, bounded operator, and that the definition does not depend on the choice of the particular basis vectors $f_0, f_1, \dots$.

**Theorem 10.4.2.** *Let $A$ be a compact normal operator on a separable Hilbert space $H$.*

1. *If $g$ is a bounded function on $\sigma_p(A)$, then $\|g(A)\| = \sup\{|g(\lambda)| : \lambda \in \sigma_p(A)\}$.*

2. *If $g$ and $h$ are bounded functions on $\sigma_p(A)$, then $(g+h)(A) = g(A)+h(A)$ and $gh(A) = g(A)h(A)$.*

3. *If $\{g_n\}$ is a sequence of functions bounded on $\sigma_p(A)$ that converge uniformly on $\sigma_p(A)$ to $g$, then $g_n(A) \to g(A)$.*

4. *If $g$ is continuous on $\sigma(A)$, then $g(A) = \lim_n g_n(A)$, where $g_n$ are polynomials.*

*Proof.* The first two items are elementary and are left as exercises for the reader. The third item then follows from the first two, since

$$\|g(A) - g_n(A)\| = \|(g - g_n)(A)\| = \sup\{|g(\lambda) - g_n(\lambda)| : \lambda \in \sigma_p(A)\} \to 0.$$

The final item follows from the third one and the Stone-Weierstrass theorem. $\square$

**Exercise 10.4.3.** Note that in the fourth item of the theorem, we asked that $g$ be continuous on $\sigma(A)$, not just on $\sigma_p(A)$. Why?

As a basic application of the functional calculus, we prove the existence of a positive square root for compact positive operators. Recall that an operator $A \in B(H)$ is said to be **positive** (written $A \geq 0$) if $\langle Ah, h \rangle \geq 0$ for all $h \in H$.

**Theorem 10.4.4.** *Let $A$ be a positive compact operator on a Hilbert space $H$. Then, there exists a unique positive operator $B$ on $H$, such that*

$$A = B^2.$$

*Moreover, $B$ is compact.*

The operator $B$ in the theorem is called the **positive square root** of $A$, and is denoted by $\sqrt{A}$ or $A^{1/2}$.

*Proof.* Since $A \geq 0$, all its eigenvalues are nonnegative (see the exercise below). Let $g : [0, \infty) \to [0, \infty)$ be given by $g(x) = \sqrt{x}$, and define $B = g(A)$. By the second item of Theorem 10.4.2, $B^2 = A$. The operator $B$ is compact because it is a diagonal operator with eigenvalues converging to 0.

As for uniqueness, suppose that $C \geq 0$ satisfies $C^2 = A$. Then $CA = C^3 = AC$. By Lemma 10.3.5, for every $\lambda \in \sigma_p(A)$, $N_\lambda$ is a reducing subspace for $C$. If $\lambda \neq 0$, then $N_\lambda$ is finite dimensional so $C\big|_{N_\lambda}$ is a compact self-adjoint operator — therefore diagonalizable — such that $(C\big|_{N_\lambda})^2 = \lambda I_\lambda$. Thus $C\big|_{N_\lambda} = \sqrt{\lambda} I_{N_\lambda} = B\big|_{N_\lambda}$.

Finally, if $0 \in \sigma_p(A)$, then $N_0$ might be infinite dimensional, so we need to argue differently. Now $D = C\big|_{N_0}$ is a self-adjoint operator such that $D^2 = 0$, so $\|D\|^2 = \|D^*D\| = \|D^2\| = 0$, meaning that $D = 0 = B\big|_{N_0}$. We conclude that $C = B$. $\qquad\square$

**Exercise 10.4.5.** Prove that for a compact self-adjoint operator $A$

$$A \geq 0 \quad \text{if and only if} \quad \sigma(A) \subseteq [0, \infty).$$

## 10.5  Additional exercises

**Exercise 10.5.1.** Let $A \in B(H)$ be a compact operator. Prove that $\ker A^\perp$ and $\overline{\operatorname{Im} A}$ are separable.

**Exercise 10.5.2.** Prove that a bounded operator $A$ on a Hilbert space is diagonal if and only if $H$ has an orthonormal basis that consists of eigenvectors of $A$. Find the relationship between the spectrum and the eigenvalues of $A$.

**Exercise 10.5.3.** In this exercise, we complete the definition of integral operators on $L^2[0,1]$ as hinted in Example 9.1.9. Recall, that in that example we claimed that given $k \in L^2\left([0,1]^2\right)$, one can define an operator $K : L^2[0,1] \to L^2[0,1]$ by

$$Kf(x) = \int_0^1 k(x,t)f(t)dt \ , \ f \in L^2[0,1]. \tag{10.5}$$

To make sense of this formula without measure theory, we proceed as follows.

1. Prove that if $k \in C\left([0,1]^2\right)$, then there exists a bounded operator $K$ on $L^2[0,1]$, such that for all $f \in C([0,1])$, $Kf$ is given by (10.5). Moreover, show that $\|K\| \leq \|k\|_2 = \left(\int_{[0,1]^2} |k(x,y)|^2 dxdy\right)^{1/2}$.

2. Prove that if $k \in C\left([0,1]^2\right)$, then the bounded operator $K$ defined above is compact. (**Hint:** prove it first for the case where $k$ is a trigonometric polynomial, i.e., $k(x,y) = \sum_{m,n=-N}^{N} a_{m,n} e^{2\pi i(mx+ny)}$.)

3. Prove that if $k \in L^2\left([0,1]^2\right)$, if $\{k_n\}$ is a sequence of continuous functions such that $k_n \xrightarrow{\|\cdot\|_2} k$, and if $K_n$ denotes the integral operator corresponding to $k_n$, then $\{K_n\}$ is a Cauchy sequence in $B(L^2[0,1])$.

Show that if $K = \lim_n K_n$, then $K$ is a well-defined compact operator on $L^2[0,1]$, and that, moreover, $K$ does not depend on the particular choice of sequence $k_n \to k$.

4. Prove that if $k \in L^2\left([0,1]^2\right)$ happens to be continuous, then the operator $K$ defined above by using Cauchy sequences coincides with the operator defined by (10.5).

**Exercise 10.5.4.** Let $k \in C\left([0,1]^2\right)$ and let $K$ be the integral operator defined by (10.5).

1. Is it possible that $\|K\| = \|k\|_2$? Does $\|K\| = \|k\|_2$ necessarily hold?

2. Compute the adjoint of $K$, and give a meaningful necessary and sufficient condition for that $K$ is self-adjoint in terms of: **(a)** the values of $k$, and **(b)** the Fourier coefficients of $k$.

**Exercise 10.5.5.** Show that for an operator $A$ on a complex Hilbert space $H$,
$$\|A\| \leq 2r(A),$$
and that if $A$ is normal, then
$$\|A\| = r(A).$$
What happens if $H$ is a real Hilbert space?

**Exercise 10.5.6.** Prove that if $A$ is a normal operator on a Hilbert space, and if $h$ is an eigenvector of $A$ corresponding to the eigenvalue $\lambda$, then $h$ is also an eigenvector for $A^*$ corresponding to an eigenvalue $\overline{\lambda}$ (in particular, the eigenvalues of a self-adjoint operator are real). Does this remain true if $A$ is not assumed normal?

**Exercise 10.5.7.** Let $f \in C_{per}([0,1])$, and let $C_f : L^2[0,1] \to L^2[0,1]$ be the associated convolution operator given by
$$C_f h(x) = \int_0^1 f(x-t)h(t)dt,$$
for all continuous $h$. Prove directly that $C_f$ is a normal operator. Characterize for which $f$ the operator $C_f$ is compact, and when this happens, find its diagonalization (i.e., find an orthonormal basis of eigenvectors and the corresponding eigenvalues).

**Exercise 10.5.8.** For an injective and positive compact operator $A \in B(H)$, let $\lambda_1(A) \geq \lambda_2(A) \geq \ldots$ denote its eigenvalues (corresponding to an orthonormal basis of eigenvectors) written in decreasing order.

1. Prove that for every $n = 1, 2, \ldots$,
$$\lambda_n(A) = \min_{\substack{M \subset H \\ \dim M = n-1}} \max_{\substack{\|h\|=1 \\ h \perp M}} \langle Ah, h \rangle.$$

2. Why did we assume that $A$ is injective? How should the formulation be modified if $A$ is not necessarily injective?

3. Show that if $A \le B$ (meaning that $B - A \ge 0$), then

$$\lambda_n(A) \le \lambda_n(B), \quad \text{for all } n = 1, 2, \ldots$$

4. Prove that

$$|\lambda_n(A) - \lambda_n(B)| \le \|A - B\|, \quad \text{for all } n = 1, 2, \ldots$$

**Exercise 10.5.9** (The maximum modulus principle). Let $z \in \mathbb{D}$, and let $f(z) = \sum_{n=0}^{\infty} a_n z^n$ be an analytic function with a power series that converges in an open neighborhood of $\overline{\mathbb{D}}$. Use the spectral theorem to prove that

$$|f(z)| \le \max_{|w|=1} |f(w)|.$$

(**Hint:** for the case where $f$ is a polynomial of degree $n$, put $s = \sqrt{1 - |z|^2}$, and consider the $(n+1) \times (n+1)$ matrix

$$U = \begin{pmatrix} z & & & s \\ s & & & -\bar{z} \\ & 1 & & \\ & & \ddots & \\ & & & 1 \end{pmatrix}.$$

Show that $U$ is unitary, evaluate the norm of $f(U)$, and show that $U^k = \begin{pmatrix} z^k & * \\ * & * \end{pmatrix}$ for all $k = 0, 1, 2, \ldots, n$, i.e., the $1, 1$ entry of the matrix $U^k$ is equal to $z^k$.)

**Exercise 10.5.10** (von Neumann's inequality in finite dimensional spaces). Let $T$ be a contraction on a finite dimensional Hilbert space $H$. Prove that for every polynomial $f(z) = \sum_{n=0}^{N} a_n z^n$,

$$\|f(T)\| \le \sup_{|w|=1} |f(w)|.$$

(**Hint:** consider the block matrix

$$U = \begin{pmatrix} T & & & (I - TT^*)^{1/2} \\ (I - T^*T)^{1/2} & & & -T^* \\ & I_H & & \\ & & \ddots & \\ & & & I_H \end{pmatrix},$$

acting on $H \oplus \cdots \oplus H$ ($n+1$ times), and argue as in the previous exercise. To prove that $U$ is unitary, you will need to justify that $T(I - T^*T)^{1/2} = (I - TT^*)^{1/2}T^*$, which is nontrivial.)

**Exercise 10.5.11.** Let $T$ be a contraction on a finite dimensional space $H$. The hint of the previous exercise shows how to construct a Hilbert space $K \supset H$ and a unitary operator $U$ on $K$, such that

$$T^k = P_H U^k\big|_H, \quad \text{for all } k = 0, 1, 2, \ldots, n.$$

1. Prove that for such a $T$ one can find a Hilbert space $K \supset H$ and a unitary operator $U$ on $K$, such that

$$T^k = P_H U^k\big|_H, \quad \text{for all } k = 0, 1, 2, \ldots.$$

2. Prove that if $T$ is not a unitary, and if there is a Hilbert space $K \supset H$ and a unitary operator $U$ on $K$, such that $T^k = P_H U^k\big|_H$ for *all* $k = 0, 1, 2, \ldots$, then $K$ must be infinite dimensional.

**Exercise 10.5.12** (von Neumann's inequality). Let $H$ be a Hilbert space, $T \in B(H)$, and let $p$ be a polynomial. Prove that

$$\|p(T)\| \leq \sup_{|z| \leq \|T\|} |p(z)|.$$

(**Hint:** If $T$ acts on a finite dimensional space, this follows from Exercise 10.5.10. If $H$ is separable, then let $\{e_i\}_{i \in \mathbb{N}}$ be an orthonormal basis, let $P_n$ be the orthogonal projection onto $\text{span}\{e_0, e_1, \ldots, e_n\}$, and consider the operators $T_n = P_n T P_n$.) Does the above inequality remain true for contractions acting on a general Banach space?

The following three exercises outline a slightly different proof of Théorem 10.3.1, in order to acquaint the reader with some tools that are useful in a more advanced study of functional analysis.

**Exercise 10.5.13.** Prove that the closed unit ball of a Hilbert space is weakly sequentially compact, meaning that for every sequence $\{h_n\}_n$ in the unit ball $H_1$, there is a subsequence $\{h_{n_k}\}_k$ and some $h \in H_1$ such that $h_{n_k} \rightharpoonup h$.

A very close result was given as Exercise 7.5.14. We note that these results are actually a special case of *Alaoglu's Theorem*, which states that for every Banach space $X$, the closed unit ball $(X^*)_1$ of the dual space is compact in the so called *weak-* topology*.

**Exercise 10.5.14.** Let $A \in B(H, K)$. Show that $A$ is compact if and only if $A$ is "weak-to-strong continuous", in the sense that, for every weakly convergent sequence $\{h_n\}$ in $H$, the sequence $\{Ah_n\}$ is strongly convergent in $K$ (**Hint:** to show that a compact operator takes weakly convergent sequences to strongly convergent sequences, it might be helpful to use Exercise A.1.14; for the converse, use Exercise 10.5.13.)

**Exercise 10.5.15.** Use the previous exercise to give an alternative proof that $A$ is compact if and only if $A^*$ is compact. (**Hint:** supposing that $A$ is compact, show that $k_n \rightharpoonup k$ in $K$ implies that $A^* k_n \to A^* k$ in $H$.)

**Exercise 10.5.16** (Hilbert-Schmidt operators). Let $H$ be a Hilbert space and let $\{e_i\}_{i \in I}$ be an orthonormal basis. An operator $T \in B(H)$ is said to be a **Hilbert-Schmidt** operator if $\sum_{i \in I} \|Te_i\|^2 < \infty$. We define

$$\|T\|_2 = \left( \sum_{i \in I} \|Te_i\|^2 \right)^{1/2}.$$

The quantity $\|\cdot\|_2$ is called the **Hilbert-Schmidt norm**. The set of all Hilbert-Schmidt operators on $H$ is denoted by $\mathcal{S}(H)$. Prove the following facts.

1. The Hilbert-Schmidt norm of an operator is independent of the basis, that is
$$\sum_{i \in I} \|Te_i\|^2 = \sum_{j \in J} \|Tf_j\|^2$$
for any orthonormal basis $\{f_j\}_{j \in J}$.

2. $\|T\|_2 = \|T^*\|_2 \geq \|T\|$ for any $T \in B(H)$.

3. For any $A \in B(H)$ and $T \in \mathcal{S}(H)$, both $AT$ and $TA$ are in $\mathcal{S}(H)$, and moreover
$$\|AT\|_2 \leq \|A\|\|T\|_2 \quad \text{and} \quad \|TA\|_2 \leq \|A\|\|T\|_2.$$

4. If $[T] = (t_{ij})$ denotes the matrix representation of $T$ with respect to any basis, then
$$\|T\|_2 = \left( \sum_{i,j} |t_{ij}|^2 \right)^{1/2}.$$

5. Every finite rank operator is Hilbert-Schmidt, and $\mathcal{S}(H) \subsetneq K(H)$.

6. $\|\cdot\|_2$ defines a norm on $\mathcal{S}(H)$, which makes $\mathcal{S}(H)$ into a Banach space.

7. The norm $\|\cdot\|_2$ is induced by an inner product. Describe the inner product which induces this norm.

8. Find an orthonormal basis for $\mathcal{S}(H)$.

**Exercise 10.5.17.** Prove that if $k \in L^2\left([0,1]^2\right)$ and $K$ is the corresponding integral operator on $L^2[0,1]$, then $K$ is a Hilbert-Schmidt operator. Show that, conversely, any Hilbert-Schmidt operator on $L^2[0,1]$ has this form.

**Exercise 10.5.18.** Let $A$ be a self-adjoint operator on a Hilbert space, and suppose that $A^2$ is compact. Prove that $A$ is compact.

# Chapter 11

## Applications of the theory of compact operators to problems in analysis

In this chapter, we will present two problems in analysis to which the theory of compact operators can be applied.

In the first section, we finally return to the integral equation that was mentioned in the introduction, and see how the theory we have learned applies. Integral equations are the classical example of a concrete analytical problem to which the theory of compact operators can be applied. Although we will obtain significant results with ease, we should modestly acknowledge that we will not obtain any result that was not already known to Hilbert and his contemporaries by 1910 (see [7, Chapter V] or Chapters 43, 45 and 46 in [17].). There are numerous monographs devoted entirely to the subject, which has seen massive development since the time of Hilbert; see, for example, the classic one by Tricomi [34].

In the second section, we treat the solvability of certain functional equations, following Paneah. This is a rather recent problem to which the theory of compact operators that we learned can be applied.

## 11.1   Integral equations

We now return to the integral equation

$$f(x) + \int_a^b k(t,x)f(t)dt = g(x),$$

which was discussed in the introduction. Let us see how the operator theory that we learned helps in understanding the solvability of such equation. We will only scratch the surface; the reader will be able to find a broader and well-motivated introductory treatment of integral operators in Gohberg and Goldberg's book [12] (it was from that book that I first learned anything about integral equations, and I also kept it handy when preparing this section).

Without loss of generality we set $[a,b] = [0,1]$, and, following Hilbert, we rewrite the integral equation as

$$f(x) - \lambda \int_0^1 k(t,x)f(t)dt = g(x), \qquad (11.1)$$

where $\lambda$ is some nonzero complex number. We assume that $k$ is continuous on $[0,1]^2$, and we define an operator $K : C([0,1]) \to C([0,1])$ by

$$Kf(x) = \int_0^1 k(t,x)f(t)dt. \qquad (11.2)$$

In Example 9.1.6 it was proved that $K$ determines a compact operator on $C([0,1])$, when this space is given the supremum norm. It is also evident that if $C([0,1])$ is endowed with the norm $\|\cdot\|_2$ (that is, if $C([0,1])$ is considered as a subspace of $L^2[0,1]$), then $K$ is a bounded operator, and therefore it extends to a bounded operator $K : L^2[0,1] \to L^2[0,1]$. In Example 9.1.9 we claimed that $K$ is a compact operator on $L^2[0,1]$ (detailed guidance is provided in Exercise 10.5.3). We will consider the equation

$$(I - \lambda K)f = g \qquad (11.3)$$

in the space $C([0,1])$ as well as in $L^2[0,1]$.

It will be fruitful to consider also Volterra type operators

$$Kf(x) = \int_0^x k(x,t)f(t)dt. \qquad (11.4)$$

In Exercise 9.3.5, the reader was asked to prove that this operator is a compact operator on $C([0,1])$. The same argument that shows that the operator given by (11.2) extends to a compact operator on $L^2[0,1]$ shows that the operator given by (11.4) extends to a compact operator on $L^2[0,1]$.

We will make no attempt to find the largest class of kernels, for which a unified treatment of operators on $C([0,1])$ and on $L^2[0,1]$ can be given. We are deliberately working with this limited class of kernels, to keep things clear and simple.

## 11.1.1   The setting of continuous functions

**Theorem 11.1.1.** *Let $k$ be a continuous function on $[0,1]^2$, and let $K : C([0,1]) \to C([0,1])$ be the operator defined either by (11.2) or by (11.4).*

1. *If $\lambda^{-1} \notin \sigma(K)$, then for every $g \in C([0,1])$, Equation (11.3) has a unique solution $f \in C([0,1])$.*

2. *In particular, (11.3) is uniquely solvable for all complex numbers, except, possibly, a sequence $\{\lambda_n\}$ that is either finite or converges to $\infty$.*

3. *A necessary and sufficient condition for there to be a unique solution $f$ for (11.3) given any $g \in C([0,1])$, is that the only solution $f$ of the homogeneous equation $(I - \lambda K)f = 0$ is the trivial solution $f = 0$.*

4. If $\lambda^{-1} \notin \sigma(K)$, then $f$ depends continuously on $g$, in the sense that if $g_n \to g$, and if $f_n$ is the solution of the equation $(I - \lambda K)f_n = g_n$, then $f_n \to f$.

5. If $\lambda^{-1} \in \sigma(K)$, then the space of solutions to $(I - \lambda K)f = 0$ is finite dimensional. If $\{f_1, \ldots, f_n\}$ is a basis for $\ker(I - \lambda K)$, and $f_0$ is a particular solution to (11.1), then the general solution of (11.3) is given by

$$f = f_0 + c_1 f_1 + \ldots + c_n f_n.$$

6. If there is some $n \in \mathbb{N}$ such that $\|(\lambda K)^n\| < 1$, then $\lambda^{-1} \notin \sigma(K)$, and therefore (11.3) is uniquely solvable for every $g$. Moreover, the solution $f$ is given by

$$f = (I - \lambda K)^{-1} g = \sum_{n=0}^{\infty} (\lambda K)^n g. \qquad (11.5)$$

In this case, the inverse of $I - \lambda K$ is given by $(I - \lambda K)^{-1} = I + \tilde{K}$, where $\tilde{K}$ is compact operator.

*Proof.* All the claims, except the last one, follow from the spectral theory of compact operators (Section 9.2).

The last claim follows from the solution of Exercise 8.5.7. To elaborate, mimicking the proof of Proposition 8.3.9, one shows that the series $\sum_{n=0}^{\infty} (\lambda K)^n$ converges to the inverse of $I - \lambda K$. The operator $\tilde{K} = \sum_{n=1}^{\infty} (\lambda K)^n$ is a compact operator such that $(I - \lambda K)^{-1} = I + \tilde{K}$. $\qquad \square$

**Example 11.1.2.** We will now show that if $K$ is a Volterra type operator, then $\|(\lambda K)^n\| \to 0$ for any $\lambda \in \mathbb{C}$, and therefore the integral equation can be solved by the Neumann series (11.5). We also show that the operator $\tilde{K}$, for which $(I - \lambda K)^{-1} = I + \tilde{K}$, is an integral operator, and give a formula for the kernel $\tilde{k}$ that gives rise to it.

First, we find a formula for the operator $K^n$, $n = 1, 2, \ldots$. If $f \in C([0, 1])$, then

$$K^2 f(x) = \int_0^x k(x, t)(Kf)(t)dt$$

$$= \int_0^x k(x, t) \left( \int_0^t k(t, s)f(s)ds \right) dt$$

$$= \int_0^x \left( \int_s^x k(x, t)k(t, s)f(s)dt \right) ds$$

$$= \int_0^x k_2(x, s)f(s)ds,$$

where $k_2(x, s) = \int_s^x k(x, t)k(t, s)dt$.

Continuing this way, we find that

$$K^n f(x) = \int_0^x k_n(x,t) f(t) dt,$$

where $k_n$ is defined recursively by

$$k_n(x,t) = \int_t^x k(x,s) k_{n-1}(s,t) ds.$$

Next, we prove that $|k_n(x,t)| \leq \frac{\|k\|_\infty^n}{(n-1)!} x^{n-1}$ for all $0 \leq t \leq x \leq 1$. This is evident for $n = 1$, and assuming that it is true for some $n \geq 1$, we compute

$$|k_{n+1}(x,t)| \leq \int_t^x |k(x,s)| |k_n(s,t)| ds$$

$$\leq \int_0^x \|k\|_\infty \frac{\|k\|_\infty^n}{(n-1)!} s^{n-1} ds$$

$$= \frac{\|k\|_\infty^{n+1}}{(n-1)!} \int_0^x s^{n-1} ds$$

$$= \frac{\|k\|_\infty^{n+1}}{n!} x^n.$$

The series $\sum_{n=1}^\infty \lambda^n k_n(x,t)$ therefore converges uniformly on $\{(x,t) : 0 \leq t \leq x \leq 1\}$ to a continuous function $\tilde{k}$ (it is not hard, but also not really necessary, to show that $\tilde{k}$ can be continued to a continuous function on all of $[0,1]^2$). Moreover,

$$\|K^n\| \leq \sup_{0 \leq t \leq x \leq 1} |k_n(x,t)|,$$

so the series $\sum_{n=1}^\infty (\lambda K)^n$ converges in norm to an operator $\tilde{K}$. It follows that

$$\tilde{K} f(x) = \int_0^x \tilde{k}(x,t) f(t) dt.$$

Thus, we have found an explicit inversion formula for the inverse of $I - \lambda K$ for any integral operator $K$ that is of Volterra type.

**Exercise 11.1.3.** Provide the details justifying that $\tilde{K}$ is the integral operator determined by $\tilde{k}$.

**Exercise 11.1.4.** Let $k \in C\left([0,1]^2\right)$ and let $K$ be given by (11.2). Find a formula for $K^n$. Show that there exist choices of $k$ for which $\|K^n\| \to \infty$.

## 11.1.2  *The setting of square integrable functions

In the previous section, we gathered some information regarding the solvability of (11.3) in the space $C([0,1])$. In some nice cases it is possible to write down a solution in terms of the Neumann series $\sum(\lambda K)^n$. However, the Neumann series does not always converge, even when $\lambda \notin \sigma(K)$. We now consider the equation (11.3) in $L^2[0, 1]$. The advantage of working in $L^2[0, 1]$, is that there are more tools that suggest themselves by which we can study the solvability and find solutions.

A first remark is that Theorem 11.1.1, as well as all the conclusions regarding Volterra type operators, holds true in this setting as well, with the same justifications.

### 11.1.2.1    A condition for solvability

By the $L^2$ version of Theorem 11.1.1, we know that the integral equation (11.3) has a solution $f$ for every $g \in L^2[0, 1]$, if and only if $\ker(I - \lambda K) = \{0\}$. We shall now present a sharper condition for solvability, making use of the additional structure available in Hilbert space.

We begin with an important lemma.

**Lemma 11.1.5.** *Let $H$ be a Hilbert space. For every $A \in K(H)$, the image $\operatorname{Im}(I - A)$ of $I - A$ is closed.*

*Proof.* Let $N = \ker(I - A)^\perp$ be the orthogonal complement of the kernel of $I - A$. Clearly, $\operatorname{Im}(I - A) = \operatorname{Im}(I - A)|_N$. However, $\ker(I - A) \cap N = \{0\}$, so by Lemma 9.2.6, $\operatorname{Im}(I - A)|_N = (I - A)N$ is closed. $\qquad\square$

We can now prove the converse of Lemma 9.2.7 in Hilbert spaces. It is worth noting that there are versions of Lemma 11.1.5 and Corollary 11.1.6 that hold true in any Banach space, but the proof in the general case requires tools that are beyond the scope of this book.

**Corollary 11.1.6.** *Let $H$ be a Hilbert space. For every $A \in K(H)$,*

$$\ker(I - A) = \{0\} \quad \text{if and only if} \quad \operatorname{Im}(I - A) = H.$$

*Proof.* By Theorem 9.2.9, $\ker(I - A) = \{0\}$ if and only if $1 \notin \sigma(A)$, and this happens (by Exercise 8.3.20) if and only if $1 = \bar{1} \notin \sigma(A^*)$. By Theorem 10.3.1, $A^*$ is also compact, so using 9.2.9 once more we see that $1 \notin \sigma(A^*)$ is equivalent to $\ker(I - A^*) = \{0\}$. By Proposition 5.4.8,

$$\overline{\operatorname{Im}(I - A)} = \ker(I - A^*)^\perp.$$

Putting everything together, we see that $\ker(I - A) = \{0\}$ if and only if $\overline{\operatorname{Im}(I - A)} = H$. But by Lemma 11.1.5, $\operatorname{Im}(I - A)$ is closed, and so we are done. $\qquad\square$

Returning to integral operators, we conclude that $\operatorname{Im}(I - \lambda K) = L^2[0, 1]$ if and only if $\ker(I - \lambda K) = \{0\}$.

The following theorem treats the case that $\ker(I - \lambda K) \neq \{0\}$ (equivalently, $\ker(I - (\lambda K)^*) \neq \{0\}$), and gives a finite set of conditions that one may check to determine, given a function $g \in L^2[0, 1]$, whether there exists a solution to the equation $f - \lambda K f = g$.

**Theorem 11.1.7.** *Let $k$ be a continuous function on $[0,1]^2$, and let $K :$ $L^2[0,1] \to L^2[0,1]$ the operator defined either by (11.2) or by (11.4). Let $\lambda$ be such that $\lambda^{-1} \in \sigma(K)$, and let $\{\varphi_1, \ldots, \varphi_n\}$ be a basis for $\ker(I - (\lambda K)^*)$. For every $g \in L^2[0,1]$, the equation $(I - \lambda K)f = g$ has a solution if and only if*

$$\int_0^1 g(t)\overline{\varphi_i(t)}dt = 0 \quad \text{for all } i = 1, \ldots, n.$$

*Proof.* This is immediate from the relation $\overline{\text{Im}(I - \lambda K)} = \ker(I - (\lambda K)^*)^\perp$ together with the fact that $\text{Im}(I - \lambda K)$ is closed. □

### 11.1.2.2   Matrix representation, truncation, and approximation

We now restrict attention to integral operators of the form (11.2), where $k \in C\left([0,1]^2\right)$. For brevity, we consider the equation $f - Kf = g$.

A very useful fact in computing the solution of the integral equation is that for every orthonormal basis $\{e_i\}_{i \in I}$ for $L^2[0,1]$, the operator $K$ has a representing matrix $[K] = (k_{ij})_{i,j \in I}$, given by

$$k_{ij} = \langle Ke_j, e_i \rangle.$$

Assume that the basis elements $\{e_i\}_{i \in I}$ are all piecewise continuous. By Exercise 3.6.16 the collection $\{f_{ij}\}_{i,j \in \mathbb{N}}$ given by $f_{ij}(x,y) = e_i(x)\overline{e_j(y)}$ is an orthonormal basis for $L^2\left([0,1]^2\right)$. Then we have

$$\langle Ke_j, e_i \rangle = \int_0^1 \left( \int_0^1 k(x,t)e_j(t)dt \right) \overline{e_i(x)}dx$$

$$= \int_{[0,1]^2} k(x,t)\overline{e_i(x)}e_j(t)dxdt$$

$$= \langle k, f_{ij} \rangle_{L^2([0,1]^2)}.$$

In other words, the matrix element $k_{ij}$ is given by the generalized Fourier coefficients (and when the basis $\{e_i\}_{i \in I}$ is taken to be the system $\{e^{2\pi inx}\}_{n \in \mathbb{Z}}$, then these are *the* Fourier coefficients of $k$). It follows from Parseval's identity that

$$\sum_{i,j \in I} |k_{ij}|^2 = \|k\|_{L^2}^2. \tag{11.6}$$

In particular, $K$ is a Hilbert-Schmidt operator (recall Exercise 10.5.16).

Thus, an integral equation of the form $(I - K)f = g$ is equivalent to a corresponding system of infinitely many equations in infinitely many unknowns. Choosing (as we may) an orthonormal basis $\{e_n\}_{n \in \mathbb{N}}$ indexed by the natural numbers, we are led to solve

$$\begin{pmatrix} 1 - k_{00} & -k_{01} & -k_{02} & \cdots \\ -k_{10} & 1 - k_{11} & -k_{12} & \cdots \\ -k_{20} & -k_{21} & 1 - k_{22} & \cdots \\ \vdots & \vdots & \vdots & \ddots \end{pmatrix} \begin{pmatrix} x_0 \\ x_1 \\ x_2 \\ \vdots \end{pmatrix} = \begin{pmatrix} y_0 \\ y_1 \\ y_2 \\ \vdots \end{pmatrix},$$

where $y_n = \langle g, e_n \rangle$ for all $n \in \mathbb{N}$ and $x = (x_n)$ is to be determined. The solution $f$ of $(I - K)f = g$ is then given by $f = \sum_{n \in \mathbb{N}} x_n e_n$.

The following is stated in terms of infinite matrices, but has immediate consequences to the integral equation that we are considering in this section. We will identify an infinite matrix with the multiplication operator that it induces on $\ell^2$. If $B$ is an $n \times n$ matrix, then we let $B \oplus 0$ denote the infinite matrix that has $B$ in the upper left block, and zeroes elsewhere. If $A = (a_{ij})_{i,j \in \mathbb{N}}$ is an infinite matrix, then we define $A^{(n)}$ to be the matrix $(a_{ij})_{i,j=0}^n$.

**Theorem 11.1.8.** *Let* $A = (a_{ij})_{i,j \in \mathbb{N}}$ *be an infinite matrix such that* $\sum_{i,j \in \mathbb{N}} |a_{ij}|^2 < \infty$. *Assume that* $I + A$ *is invertible, and also that for every* $n \in \mathbb{N}$, *the matrix*

$$I_{n+1} + A^{(n)} = \begin{pmatrix} 1 + a_{00} & a_{01} & \cdots & a_{0n} \\ a_{10} & 1 + a_{11} & \cdots & a_{1n} \\ \vdots & \vdots & \ddots & \vdots \\ a_{n0} & a_{n1} & \cdots & 1 + a_{nn} \end{pmatrix}$$

*is invertible. Then*

$$\|(I + A)^{-1} - (I + [A^{(n)} \oplus 0])^{-1}\| \xrightarrow{n \to \infty} 0.$$

*Proof.* By the norm estimate in (5.7),

$$\|A - [A^{(n)} \oplus 0]\|^2 \le \sum_{i>n \text{ or } j>n} |a_{ij}|^2 \to 0.$$

Therefore $\|(I + A) - (I + [A^{(n)} \oplus 0])\| \to 0$, so by Exercise 8.3.14 it follows that $\|(I + A)^{-1} - (I + [A^{(n)} \oplus 0])^{-1}\| \to 0$. $\qquad\square$

Let us spell out how the theorem can be used to find approximate solutions of the equation

$$x + Ax = y.$$

Given $y = (y_k) \in \ell^2$, the assumption is that there exists a unique solution $x = (x_k) \in \ell^2$. For every $n \in \mathbb{N}$, we put

$$X^{(n)} = \begin{pmatrix} X_0^{(n)} \\ X_1^{(n)} \\ \vdots \\ X_n^{(n)} \end{pmatrix} = \begin{pmatrix} 1 + a_{00} & a_{01} & \cdots & a_{0n} \\ a_{10} & 1 + a_{11} & \cdots & a_{1n} \\ \vdots & \vdots & \ddots & \vdots \\ a_{n0} & a_{21} & \cdots & 1 + a_{nn} \end{pmatrix}^{-1} \begin{pmatrix} y_0 \\ y_1 \\ \vdots \\ y_n \end{pmatrix}.$$

Now let

$$x^{(n)} = \begin{pmatrix} X_0^{(n)} \\ \vdots \\ X_n^{(n)} \\ 0 \\ 0 \\ \vdots \end{pmatrix}.$$

We claim that the sequence $\{x^{(n)}\}$ converges in norm to the true solution of the system, that is $x^{(n)} \to x$. From this, it also follows that the $x^{(n)}$ approximately solve the equation, since $(I + A)x^{(n)} \to (I + A)x = y$. Put

$$\tilde{x}^{(n)} = \begin{pmatrix} X_0^{(n)} \\ \vdots \\ X_n^{(n)} \\ y_{n+1} \\ y_{n+2} \\ \vdots \end{pmatrix}.$$

Note that $\tilde{x}^{(n)} = (I + [A^{(n)} \oplus 0])^{-1}y$. Therefore, by the theorem, $\tilde{x}^{(n)} \to (I + A)^{-1}y = x$. On the other hand, $\|\tilde{x}^{(n)} - x^{(n)}\|^2 = \sum_{k=n+1}^{\infty} |y_k|^2 \to 0$. It follows that $x^{(n)} \to x$, as we claimed.

We conclude this section with two classes of examples to which the above procedure applies.

**Example 11.1.9.** Suppose that $A = (a_{ij})$ satisfies $\sum_{ij} |a_{ij}|^2 < 1$. Then the assumptions of the theorem are satisfied, thanks to Proposition 8.3.9. In this situation the solution $x$ can be found by using the Neumann series, but note that computing $\sum_{n=0}^{N} A^n y$ involves taking the powers of an infinite matrix, while computing the appoximants $x^{(n)}$ involves operating with finite matrices only.

**Example 11.1.10.** Suppose that $A$ is a Hilbert Schmidt matrix such that $A \geq 0$. Then $I + A$ is a diagonalizable operator, with all eigenvalues in $[1, \infty)$, and therefore $I + A$ is invertible. Likewise $A^{(n)} \geq 0$ for all $n$, so $I_{n+1} + A^{(n)}$ is invertible. Therefore, the above method applies.

### 11.1.2.3 Normal integral operators

**Proposition 11.1.11.** *Let $A$ be a compact normal operator on a Hilbert space $H$. Let $\{e_i\}_{i \in I}$ be an orthonormal basis for $H$, and $\{\alpha_i\}_{i \in I}$ a family of complex scalars convergent to 0, such that*

$$Ah = \sum_{i \in I} \alpha_i \langle h, e_i \rangle e_i,$$

*for all $h \in H$. Then, for every nonzero $\lambda$ such that $\lambda^{-1} \notin \sigma(A)$ and all $h \in H$,*

$$(I - \lambda A)^{-1}h = \sum_{i \in I} (1 - \lambda \alpha_i)^{-1} \langle h, e_i \rangle e_i$$

$$= h + \sum_{i \in I} \frac{\lambda \alpha_i}{1 - \lambda \alpha_i} \langle h, e_i \rangle e_i,$$

*In other words,*

$$(I - \lambda A)^{-1} = I + B,$$

*where B is the compact normal operator given by*

$$Bh = \sum_{i \in I} \frac{\lambda \alpha_i}{1 - \lambda \alpha_i} \langle h, e_i \rangle e_i.$$

*Proof.* This proposition is about inverting a diagonal operator — a straightforward matter. Keeping in mind that $|1 - \lambda \alpha_i|$ is bounded below, the reader will have no problem to supply a proof. ☐

Now let $K$ be a normal integral operator on $L^2[0,1]$, given by (11.2), where $k \in C\left([0,1]^2\right)$. Then, by Theorem 10.2.11 and Exercise 10.2.4, there exists an orthonormal basis $\{e_n\}_{n \in \mathbb{N}}$ corresponding to a family of eigenvalues $\{\lambda_n\}_{n \in \mathbb{N}}$, such that $K$ is given by the convergent sum

$$K = \sum_{n \in \mathbb{N}} \lambda_n e_n \otimes e_n^*.$$

Therefore, if we are able to find the basis that diagonalizes the integral operator, then the above proposition tells us how to solve (11.3) when $\lambda^{-1} \notin \sigma(K)$.

**Example 11.1.12.** Suppose that $\varphi \in C_{per}([0,1])$ and let $k(s,t) = \varphi(s - t)$ (where as usual, we let $\varphi(x) = \varphi(x+1)$ for $x \in [-1,0)$). In Example 5.6.3, it was shown that the corresponding integral operator $K$ (which is also the convolution operator $C_\varphi$) is diagonalizable, where the orthonormal basis consisting of eigenvectors is $\{e^{2\pi i n x}\}_{n \in \mathbb{Z}}$, and the corresponding eigenvectors are $\{\hat{\varphi}(n)\}_{n \in \mathbb{Z}}$.

By the proposition, we have that for all $\lambda$ such that $\lambda^{-1} \notin \sigma_p(K) = \{\hat{\varphi}(n)\}_{n \in \mathbb{Z}}$, the operator $I - \lambda K$ is invertible, and the unique solution $f$ to (11.3) is determined by its Fourier coefficients, given by

$$\hat{f}(n) = (1 - \lambda \hat{\varphi}(n))^{-1} \hat{g}(n) = \hat{g}(n) + \frac{\lambda \hat{\varphi}(n)}{1 - \lambda \hat{\varphi}(n)} \hat{g}(n).$$

**Exercise 11.1.13.** Let $k \in C\left([0,1]^2\right)$, and let $K$ be the corresponding integral operator given by (11.2). Prove $K^*$ is the integral operator corresponding to the kernel $\tilde{k}(s,t) = \overline{k(t,s)}$. Conclude that $K$ is self-adjoint if and only if

$$\overline{k(s,t)} = k(t,s) \quad \text{for all } s,t \in [0,1].$$

such a kernel is said to be **Hermitian.**

**Theorem 11.1.14.** *Let $k \in C\left([0,1]^2\right)$ be a Hermitian kernel. Suppose that $K = \sum_{i \in I} \lambda_i e_i \otimes e_i^*$, where $\{e_i\}_{i \in I}$ is an orthonormal basis consisting of eigenvectors, corresponding to real eigenvalues $\{\lambda_i\}_{i \in I}$. Then*

$$\sum_{i \in I} \lambda_i^2 = \int_{[0,1]^2} |k(x,y)|^2 dx dy.$$

*Proof.* The representing matrix of $K$ with respect to $\{e_i\}_{i \in I}$ is diagonal with $\lambda_i$ on the diagonal, so the result follows from (11.6). ☐

### 11.1.3   *Transferring results from the square integrable setting to the continuous setting

Suppose that we are given $k \in C\left([0,1]^2\right)$, $g \in C([0,1])$, and that we are interested in studying the integral equation (11.1) in the realm of continuous functions. In the previous section we saw that, when working in $L^2[0,1]$, we can say more about when there exists a solution, and how to obtain a solution in $L^2[0,1]$ when one exists. However, we may be interested in continuous solutions only. Luckily, when the given $k$ and $g$ are continuous, then the solution $f$ is continuous, and the conditions for checking whether the problem is solvable also involve continuous functions.

Let $K : C([0,1]) \to C([0,1])$ be the integral operator corresponding to $k$, given by (11.2) or by (11.4). To avoid confusion, let $\widehat{K}$ denote the extension of $K$ to $L^2[0,1]$.

First, if $(I - \lambda\widehat{K})f = g$, then $f \in C([0,1])$, because, by

$$f(x) = \lambda \int_0^1 k(x,t)f(t)dt + g(x),$$

$f$ is manifestly continuous (the same holds if $K$ is a Volterra type operator). To be a little more precise, we argue as follows. Let $T : (C([0,1]), \|\cdot\|_2) \to (C([0,1]), \|\cdot\|_\infty)$ be the operator given by $T = \widehat{K}\big|_{C([0,1])}$. Of course, $T = K$, but we use a different letter to denote them because $T$ maps between different normed spaces. Now, $T$ is bounded:

$$\|Tf\|_\infty = \sup_{x\in[0,1]} \left| \int_0^1 k(x,t)f(t)dt \right| \leq \sup_{x\in[0,1]} \left( \int_0^1 |k(x,t)|^2 dt \right)^{1/2} \|f\|_2.$$

It follows that $T$ extends uniquely to a continuous operator from $L^2[0,1]$ into $(C([0,1]), \|\cdot\|_\infty)$. This extension is equal to $\widehat{K}$. Therefore, $f = \widehat{K}f + g$ implies that $f \in C([0,1])$.

Taking $g = 0$, we find also that every $f \in \ker(I - \lambda\widehat{K})$ is continuous, therefore $\ker(I - \lambda K) = \ker(I - \lambda\widehat{K})$. Likewise, if $\varphi \in \ker(I - (\lambda\widehat{K})^*)$, then $\varphi$ is continuous.

We conclude that if we are given continuous functions $k$ and $g$, we may work in $L^2[0,1]$, using Hilbert theoretic methods, to find the solution that we seek in $C([0,1])$. Moreover, if we want to check whether for a certain $g$ there exists a solution, we may use the condition given in Theorem 11.1.7:

$$g \in \operatorname{Im}(I - \lambda K) \quad \text{if and only if } g \perp \varphi_i, \ i = 1, \ldots, n,$$

where $\varphi_1, \ldots, \varphi_n$ is a basis (consisting of continuous functions) for $\ker(I - (\lambda K)^*)$.

The most important part of the above discussion is summarized in the following theorem, for which we give a more conceptual proof.

**Proposition 11.1.15.** *Let $k$ be a continuous function on $[0,1]^2$, and let $K :$ $C([0,1]) \to C([0,1])$ be the operator defined either by (11.2) or by (11.4). If $\widehat{K}$ is the extension of $K$ to $L^2[0,1]$, then*

$$\sigma(K) = \sigma(\widehat{K}).$$

*Moreover, if $\lambda^{-1} \notin \sigma(K)$, then*

$$(I - \lambda\widehat{K})^{-1}\big|_{C([0,1])} = (I - \lambda K)^{-1}.$$

*Proof.* The point 0 belongs to both $\sigma(K)$ and $\sigma(\widehat{K})$, as these are compact operators on infinite dimensional spaces. Thus, we need to check that $\mu \neq 0$ is an eigenvalue for $K$ if and only if it is an eigenvalue for $\widehat{K}$.

If $\mu$ is an eigenvalue for $K$, then it is clearly an eigenvalue for $\widehat{K}$ (with the same eigenvector). Now suppose that $\mu \neq 0$ is not in $\sigma_p(K)$. Then, by the Fredholm alternative (Theorem 9.2.9), $\text{Im}(\mu - K) = \text{Im}(I - \mu^{-1}K) = C([0,1])$. It follows that $\text{Im}(I - \mu^{-1}\widehat{K})$ is dense in $L^2[0,1]$, and because the image of such an operator is closed, we have $\text{Im}(I - \mu^{-1}\widehat{K}) = L^2[0,1]$. By Corollary 11.1.6 (together with the Fredholm alternative once more), $\mu \notin \sigma(\widehat{K})$.

If $g$ is continuous and $f$ is the unique continuous solution of the equation $f - \lambda K f = g$, then it must also be the unique solution in $L^2[0,1]$ to the equation $f - \lambda\widehat{K}f = g$. $\qquad\square$

---

## 11.2  *Functional equations

In this section, we will study the solvability of the functional equation

$$f(t) - f(\delta_1(t)) - f(\delta_2(t)) = h(t) , \ t \in [-1,1]. \qquad (11.7)$$

Here, $\delta_1, \delta_2$ and $h$ are given functions (satisfying some conditions, to be specified below), and the goal is to find $f$ that satisfies this equation, or to find conditions under which a solution exists. Functional equations of this type were studied by Paneah around the turn of the century, and were shown to have direct applications to problems in integral geometry and in partial differential equations (see [21, 22, 23] and the references therein). We will treat a simplified version of the problem treated by Paneah, to illustrate how the theory of compact operators has significant consequences even for problems in which there does not seem to be any compact operator in sight.

The functional equation (11.7) arose from a third-order partial differential equation. We will not derive (11.7) from the PDE (see the given references for that), we will just describe what are the natural conditions to impose on the given functions $\delta_1, \delta_2, h$ appearing in (11.7), in the context in which the functional equation arose.

First, let us work in the real-valued setting: all functions appearing below are real-valued functions. Next, we assume that the given functions $\delta_1, \delta_2, h \in C^2\left([-1,1]\right)$, where $C^2([-1,1])$ denotes the space of twice continuously differentiable functions on $[-1,1]$. We seek a solution $f \in C^2\left([-1,1]\right)$. Finally, the functions $\delta_1, \delta_2$ are assumed to satisfy:

1. $\delta_i : [-1,1] \to [-1,1]$, $i = 1,2$,

2. $\delta_1(-1) = -1$, $\delta_1(1) = \delta_2(-1) = 0$, $\delta_2(1) = 1$,

3. $\delta_i'(t) > 0$ for all $t \in [-1,1]$,

4. $\delta_1(t) + \delta_2(t) = t$ for all $t \in [-1,1]$.

With these assumptions we turn to the solution of (11.7). Since this is a linear equation, our experience in linear equations and ODEs leads us to consider first the homogeneous equation

$$f(t) - f(\delta_1(t)) - f(\delta_2(t)) = 0 , \ t \in [-1,1]. \tag{11.8}$$

The condition $\delta_1(t) + \delta_2(t) = t$ implies that $g_a(t) = at$ is a solution of the homogeneous equation for all $a \in \mathbb{R}$. It follows that if $f$ is a solution of (11.7), then $f + g_a$ is also a solution. Therefore, we may content ourselves with finding a solution $f$ for (11.7) that has $f'(0) = 0$.

Plugging $t = \pm 1$ into (11.7), we find that

$$h(-1) = -f(0) = h(1).$$

In other words, a necessary condition for (11.7) to have a solution is that $h(-1) = h(1)$, and in this case the solution $f$ must satisfy $f(0) = -h(1)$. We will restrict our attention to solving the equation for $h$ satisfying $h(-1) = h(1) = 0$, and therefore we will also look for solutions $f$ for which $f(0) = 0$. The general case is then easily taken care of (see Exercise 11.3.8).

**Theorem 11.2.1.** *For every $h \in C^2([-1,1])$ satisfying $h(-1) = h(1) = 0$, there exists a unique $f \in C^2([-1,1])$, satisfying $f(0) = f'(0) = 0$, which solves Equation* (11.7).

*Proof.* The proof goes down a long and winding road. The readers will be rewarded for their patience.

We define the following spaces:

$$X_0 = \left\{ f \in C^2([-1,1]) : f(0) = f'(0) = 0 \right\},$$

$$Y_0 = \left\{ h \in C^2([-1,1]) : h(-1) = h(1) = 0 \right\}.$$

If we endow $C^2([a,b])$ with the norm $\|g\|_{C^2} = \|g\|_\infty + \|g'\|_\infty + \|g''\|_\infty$, then it becomes a Banach space (Exercise 7.5.3), and one checks that $X_0$ and $Y_0$ are closed subspaces of $C^2([a,b])$, hence Banach spaces. Define $T_0 \in B(X_0, Y_0)$ by

$$T_0 f(t) = f(t) - f(\delta_1(t)) - f(\delta_2(t)).$$

We will prove that $T_0$ is an invertible operator (this will prove slightly more than what is stated in the theorem — it will also show that the solution $f$ depends continuously on $h$, because $f = T_0^{-1}h$ and $T_0^{-1} \in B(Y_0, X_0)$).

Differentiating (11.7), we get the equation

$$f'(t) - \delta_1'(t)f'(\delta_1(t)) - \delta_2'(t)f'(\delta_2(t)) = h'(t) , \ t \in [-1,1]. \tag{11.9}$$

In order to treat (11.9) in an operator theoretic framework, we define spaces

$$X_1 = \left\{ u \in C^1([-1,1]) : u(0) = 0 \right\},$$

$$Y_1 = \left\{ v \in C^1([-1,1]) : \int_{-1}^{1} v(t)dt = 0 \right\},$$

with norm $\|g\|_{C^1} = \|g\|_\infty + \|g'\|_\infty$. Let $D$ denote the differentiation operator $Df = f'$ defined on $C^1([-1,1])$, and let

$$D_X = D\big|_{X_0} \quad \text{and} \quad D_Y = D\big|_{Y_0}.$$

**Exercise 11.2.2.** Prove that $D_X$ is a bounded invertible map from $X_0$ onto $X_1$, and that $D_Y$ is a bounded invertible map from $Y_0$ onto $Y_1$.

The fact that differentiation of (11.7) gives (11.9) can be written compactly as

$$D_Y T_0 = T_1 D_X, \tag{11.10}$$

where $T_1 : X_1 \to Y_1$ is given by

$$T_1 g(t) = g(t) - \delta_1'(t)g(\delta_1(t)) - \delta_2'(t)g(\delta_2(t)).$$

**Exercise 11.2.3.** Use the relation (11.10) to prove that $T_0$ is injective/surjective/invertible if and only if $T_1$ is.

The following shows that $T_1$ is injective.

**Lemma 11.2.4.** *Let* $g \in C([-1,1])$ *such that*

$$g(t) = \delta_1'(t)g(\delta_1(t)) + \delta_2'(t)g(\delta_2(t)) , \ t \in [-1,1]. \tag{11.11}$$

*Then $g$ is constant.*

*Proof.* Let $t_0 \in [-1,1]$ be a point where $g$ attains its maximum. Since $\delta_1'(t_0), \delta_2'(t_0)$ are positive numbers such that $\delta_1'(t_0) + \delta_2'(t_0) = 1$, the equation (11.11) at the point $t_0$ forces $g(\delta_1(t_0)) = g(\delta_2(t_0)) = g(t_0)$. Denoting $t_1 = \delta_2(t_0)$, we get that $g(t_1) = \max_{t \in [-1,1]} g(t)$.

Arguing the same way, we find that $g$ must attain its maximum also at the point $t_2 = \delta_2(t_1)$, and if we continue in this fashion we obtain an increasing sequence of points $t_n = \delta_2(t_{n-1})$ such that $g(t_n) = \max_{t \in [-1,1]} g(t)$ for all $n$. The sequence $t_n$ converges, and the assumptions on $\delta_2$ imply that $t_n \to 1$. Conclusion: $g(1) = \max_{t \in [-1,1]} g(t)$.

But starting from a point where $g$ attains its minimum, the same argument shows that $g(1) = \min_{t \in [-1,1]} g(t)$. Therefore, $\min_{t \in [-1,1]} g(t) = \max_{t \in [-1,1]} g(t)$, whence $g$ is constant. $\square$

Differentiating (11.9) we obtain

$$f''(t) - \delta_1''(t)f'(\delta_1(t)) - \delta_1'(t)^2 f''(\delta_1(t)) - \delta_2''(t)f'(\delta_2(t)) - \delta_2'(t)^2 f''(\delta_2(t))$$
$$= h''(t),$$

which we write as $T_2 f'' = h''$ where $T_2 : C([-1,1]) \to C([-1,1])$ is the operator given by

$$T_2 k(t) = k(t) - \delta_1'(t)^2 k(\delta_1(t)) - \delta_2'(t)^2 k(\delta_2(t))$$
$$- \delta_1''(t) \int_0^{\delta_1(t)} k(s)ds - \delta_2''(t) \int_0^{\delta_2(t)} k(s)ds.$$

Note that we can write $T_2 = I - B - C$, where

$$Bk(t) = (\delta_1'(t))^2 k(\delta_1(t)) + (\delta_2'(t))^2 k(\delta_2(t)),$$

and

$$Ck(t) = \delta_1'(t) \int_0^{\delta_1(t)} k(s)ds + \delta_2'(t) \int_0^{\delta_2(t)} k(s)ds.$$

**Exercise 11.2.5.** Prove that $\|B\| < 1$ (hence $I - B$ is invertible) and that $C$ is compact.

Letting $D_X' : X_1 \to C([-1,1])$ and $D_Y' : Y_1 \to C([-1,1])$ be the differentiation operators, we have that

$$D_Y' T_1 = T_2 D_X'.$$

**Exercise 11.2.6.** Prove that $D_X'$ is a bounded invertible map from $X_1$ onto $C([-1,1])$, and that $D_Y'$ is a bounded invertible map from $Y_1$ onto $C([-1,1])$.

As in Exercise 11.2.3, one can show that $T_2$ is injective/surjective/invertible if and only if $T_1$ is.

Now comes the punch line of the proof. To show that $T_0$ is invertible, it suffices to show that $T_2$ is invertible. Now, $T_2$ has the form $I - B - C$, where $I - B$ is invertible and $C$ is compact. Therefore, $T_2$ is of the form "invertible plus compact", so Corollary 9.2.10 applies. Thus, to show that $T_2$ is invertible, it suffices to show that it is injective.

But to show that $T_2$ is injective, it suffices to show that either $T_0$ or $T_1$ is injective. By Lemma 11.2.4, $T_1$ is injective, and this completes the proof of the theorem. □

## 11.3 Additional exercises

**Exercise 11.3.1.** Let $k \in L^2([0,1]^2)$, and let $K$ be the corresponding integral operator. Prove that $K$ is diagonal with respect to $\{e^{2\pi i n x}\}_{n \in \mathbb{Z}}$ if and only if

$K$ is in fact a convolution operator, that is, there exists some $h \in L^2[0, 1]$ such that $K = C_h$ (and it follows that $k(s, t) = h(s - t)$ as elements of $L^2([0, 1]^2)$).

**Exercise 11.3.2.** Consider the statement of Theorem 11.1.8, without the assumption that $I + A$ is invertible. Does the conclusion of the theorem follow? In particular, does it follow that $I + A$ is invertible?

**Exercise 11.3.3.** Let $F = \sum_{i=1}^{n} k_i \otimes h_i^*$ be a finite rank operator on $H$, where $h_1, \ldots, h_n$ and $k_1, \ldots, k_n$ are in $H$. Prove that $I - F$ is invertible if and only if the matrix

$$\begin{pmatrix} 1 - \langle k_1, h_1 \rangle & -\langle k_2, h_1 \rangle & -\langle k_3, h_1 \rangle & \cdots & -\langle k_n, h_1 \rangle \\ -\langle k_1, h_2 \rangle & 1 - \langle k_2, h_2 \rangle & -\langle k_3, h_2 \rangle & \cdots & -\langle k_n, h_2 \rangle \\ -\langle k_1, h_3 \rangle & -\langle k_2, h_3 \rangle & 1 - \langle k_3, h_3 \rangle & \cdots & -\langle k_n, h_3 \rangle \\ \vdots & \vdots & \vdots & \ddots & \vdots \\ -\langle k_1, h_n \rangle & -\langle k_2, h_n \rangle & -\langle k_3, h_n \rangle & \cdots & 1 - \langle k_n, h_n \rangle \end{pmatrix}$$

is invertible. When this matrix is invertible, prove that the unique solution to the equation $(I - F)f = g$ is given by $f = g + \sum_{i=1}^{n} x_i k_i$, where $(x_i)_{i=1}^{n}$ is determined by

$$\begin{pmatrix} x_1 \\ x_2 \\ \vdots \\ x_n \end{pmatrix} = \begin{pmatrix} 1 - \langle k_1, h_1 \rangle & -\langle k_2, h_1 \rangle & \cdots & -\langle k_n, h_1 \rangle \\ -\langle k_1, h_2 \rangle & 1 - \langle k_2, h_2 \rangle & \cdots & -\langle k_n, h_2 \rangle \\ \vdots & \vdots & \ddots & \vdots \\ -\langle k_1, h_n \rangle & -\langle k_2, h_n \rangle & \cdots & 1 - \langle k_n, h_n \rangle \end{pmatrix}^{-1} \begin{pmatrix} \langle g, h_1 \rangle \\ \langle g, h_2 \rangle \\ \vdots \\ \langle g, h_n \rangle \end{pmatrix}.$$

Use this to find the solution to the integral equation

$$f(t) - \int_0^{2\pi} f(s) \sin(s + t) ds = e^t, \quad t \in [0, 2\pi].$$

(**Hint:** if stuck, see [12, pp. 66–69].)

**Exercise 11.3.4.** Let

$$k(s, t) = \begin{cases} (1 - s)t & 0 \le t \le s \le 1 \\ (1 - t)s & 0 \le s \le t \le 1 \end{cases},$$

and let $K$ be the corresponding integral operator on $L^2[0, 1]$,

$$K(f)(x) = \int_0^1 k(x, t) f(t) dt.$$

$K$ is a self-adjoint compact operator. Find its spectrum and find an orthonormal basis of eigenvectors. Discuss the solvability of the equations:

$$f - \lambda K f = g$$

and
$$Kf = g$$
in $L^2[0,1]$. (**Hint:** the system $\{\sin(\pi mt)\sin(\pi nt)\}_{m,n=1}^\infty$ is an orthonormal basis for $L^2([0,1]^2)$. The next exercise contains another approach for the solution of this problem.)

**Exercise 11.3.5.** This exercise (which is inspired by [12, Section IV.5]) outlines a different approach for finding the eigenvalues and eigenvectors of the operator $K$ from the previous exercise. Show that $K$ can be represented by

$$Kf(x) = -\int_0^x \int_0^t f(s)\,ds\,dt + x \int_0^1 \int_0^t f(s)\,ds\,dt,$$

for $f \in C([0,1])$. Use this to show that if $f \in C^2([0,1])$ is an eigenvector, then it satisfies a certain differential equation with boundary conditions. Even if you do not know any differential equations, you can guess the solutions. After finding enough eigenvectors in $C^2([0,1])$, explain why there are no other eigenvectors.

**Exercise 11.3.6.** By Theorem 11.1.1 (and its $L^2$ analogue) we know that if the integral equation (11.1) has a unique solution, then the unique solution $f$ depends continuously on $g$. Formulate and prove a theorem showing that $f$ depends continuously also on $k$.

**Exercise 11.3.7.** Let $A : C^2([-1,1]) \to C^2([-1,1])$ be the operator $Af = f \circ \delta_1 + f \circ \delta_2$, where $\delta_1, \delta_2$ are as in Section 11.2.

1. Compute $\|A\|$.

2. Is $A$ a compact operator?

**Exercise 11.3.8.** Prove that for every $h \in C^2([-1,1])$ such that $h(-1) = h(1)$, and every $a, b \in \mathbb{R}$, there exists a unique $f \in C^2([-1,1])$ that satisfies $f(0) = a$, $f'(0) = b$, and $f - f \circ \delta_1 - f \circ \delta_2 = h$.

**Exercise 11.3.9.** In Section 11.2, we proved that the only continuously differentiable solutions to the homogeneous functional equation (11.8) are linear functions of the form $f(t) = at$.

1. Determine if there exist continuous nonlinear solutions to the equation

$$f(t) - f\left(\frac{t-1}{2}\right) - f\left(\frac{t+1}{2}\right) = 0, \ t \in [-1,1].$$

2. Suppose that $\delta_1, \delta_2$ satisfy the conditions stated in Section 11.2. Determine if and when there exist continuous nonlinear solutions to the equation

$$f(t) - f(\delta_1(t)) - f(\delta_2(t)) = 0, \ t \in [-1,1].$$

(**Hint:** for the first question, one may use Fourier series. The second question is harder. One may consult [32].)

# Chapter 12

## The Fourier transform

If a function $f : \mathbb{R} \to \mathbb{C}$ is periodic with period $2T$, then we can study it using the associated Fourier series

$$f(t) \sim \sum_{n=-\infty}^{\infty} c_n e^{\frac{n\pi it}{T}},$$

where $c_n = \frac{1}{2T} \int_{-T}^{T} f(t) e^{\frac{-n\pi it}{T}} dt$. If $f$ is not periodic (for example, if $f$ tends to 0 at infinity), then a Fourier series cannot converge to it in any reasonable sense. In this chapter, we study a relative of Fourier series, which involves the reconstruction of $f$ using the exponential functions $\{e^{iwt}\}_{w \in \mathbb{R}}$.

The **Fourier transform** of a function $f : \mathbb{R} \to \mathbb{C}$ is defined to be the function $\hat{f} : \mathbb{R} \to \mathbb{C}$ given by

$$\hat{f}(w) = \int_{-\infty}^{\infty} f(t) e^{-iwt} dt, \tag{12.1}$$

whenever the above integral makes some sense (precise definitions will be given below).

In favorable circumstances, for example when $f$ is very smooth and tends to 0 very fast (for example, if $f(t) = e^{-t^2}$), then one can reconstruct $f$ from $\hat{f}$ by the inversion formula

$$f(t) = \frac{1}{2\pi} \int_{-\infty}^{\infty} \hat{f}(w) e^{iwt} dw. \tag{12.2}$$

The Fourier transform has countless applications in analysis, in engineering and in science. We will develop the Fourier transform for $L^1(\mathbb{R})$ and for $L^2(\mathbb{R})$ by a mixture of classical analytical and functional analytical methods. We begin with a careful definition of these spaces.

## 12.1   The spaces $L^p(\mathbb{R})$, $p \in [1, \infty)$

In this section, we will define the Banach spaces $L^p(\mathbb{R})$, for $p \in [1, \infty)$. These spaces are indispensable for doing analysis on the real line. We will first give the definition for $p = 2$, and then generalize to other values $p \in [1, \infty)$. As in our definition of the spaces $L^2[a,b]$ in Section 2.4, we will not give the standard, measure theoretic definition. Our approach suffices for the development of the basic theory of the Fourier transform. The reader can find a complete treatment of $L^p$ spaces in [11] or [29].

### 12.1.1   Definition of $L^2(\mathbb{R})$ and basic properties

Let $PC_c(\mathbb{R})$ denote the space of piecewise continuous functions on $\mathbb{R}$ which have compact support. Thus, $f : \mathbb{R} \to \mathbb{C}$ is in $PC_c(\mathbb{R})$ if there is some finite interval $[a,b]$ such that $f$ vanishes off $[a,b]$, and $f|_{[a,b]} \in PC[a,b]$. We can consider any $f \in PC[a,b]$ as a function in $PC_c(\mathbb{R})$ by simply defining it to be 0 outside of $[a,b]$. We thus identify $PC[a,b]$ as a subspace of $PC_c(\mathbb{R})$, and we have

$$PC_c(\mathbb{R}) = \bigcup_{a<b} PC[a,b]. \tag{12.3}$$

We endow $PC_c(\mathbb{R})$ with the inner product

$$\langle f, g \rangle = \int_{-\infty}^{\infty} f(t)\overline{g}(t)dt. \tag{12.4}$$

Equation (12.4) does not define a true inner product on $PC_c(\mathbb{R})$, but we solve this problem as we did in the case of $PC[a,b]$: we redefine equality, saying that $f = g$ in $PC_c(\mathbb{R})$ if and only if they are equal at all points except finitely many.

**Definition 12.1.1.** The Hilbert space $L^2(\mathbb{R})$ is defined to be the completion of $PC_c(\mathbb{R})$ with respect to the inner product (12.4).

**Proposition 12.1.2.** *For every $a < b$, the inclusion mapping $PC[a,b] \to PC_c(\mathbb{R})$ extends to an isometry $v_{[a,b]} : L^2[a,b] \to L^2(\mathbb{R})$, giving rise to an identification of $L^2[a,b]$ as a closed subspace of $L^2(\mathbb{R})$. The orthogonal projection of $L^2(\mathbb{R})$ onto $L^2[a,b]$ is the unique extension to $L^2(\mathbb{R})$ of the restriction mapping*

$$PC_c(\mathbb{R}) \to PC[a,b] \ , \ f \mapsto f|_{[a,b]}.$$

*Proof.* We know that a densely defined isometry extends uniquely to an isometry acting on the entire space. Thus, such an isometry $v_{[a,b]} : L^2[a,b] \to L^2(\mathbb{R})$ exists. The range of any isometry is closed, so $v_{[a,b]}(L^2[a,b])$ is a closed subspace of $L^2(\mathbb{R})$. From the characterization of the best approximation projection (Corollary 3.2.7), one sees that the projection $P_{L^2[a,b]}$ to $PC_c(\mathbb{R})$ is given by restriction. $\square$

We consider $L^2[a,b]$ to be that subspace of $L^2(\mathbb{R})$ that consists of functions that vanish outside of the interval $[a,b]$.

## 12.1.2   The Banach spaces $L^p(\mathbb{R})$ $(p \in [1, \infty))$

For $f \in PC_c(\mathbb{R})$, we define the so-called $L^p$ *norm* $\|\cdot\|_p$ to be

$$\|f\|_p = \left( \int_{-\infty}^{\infty} |f(t)|^p \right)^{1/p}.$$

**Definition 12.1.3.** Let $p \in [1, \infty)$. The Banach space $L^p[a,b]$ is defined to be the completion of the normed space $(PC_c([a,b]), \|\cdot\|_p)$. The Banach space $L^p(\mathbb{R})$ is defined to be the completion of the normed space $(PC_c(\mathbb{R}), \|\cdot\|_p)$.

Let $\mathcal{S}$ be the subspace of $(PC_c(\mathbb{R}), \|\cdot\|_p)$ consisting of step functions, i.e., functions of the form

$$\sum_{k=1}^{n} c_k \chi_{[a_k, b_k]}.$$

We shall require the following lemma several times below.

**Lemma 12.1.4.** *For all $p \in [1, \infty)$, $\mathcal{S}$ is dense in $(PC_c(\mathbb{R}), \|\cdot\|_p)$ in the $L^p$ norm. Therefore, for all $p \in [1, \infty)$, $\mathcal{S}$ is dense in $L^p(\mathbb{R})$*

**Exercise 12.1.5.** Prove Lemma 12.1.4.

The case $p = 1$ is of particular interest, and has one special feature which is that the mapping

$$\mathcal{I} : L^1(\mathbb{R}) \ni f \mapsto \int_{-\infty}^{\infty} f(t)dt$$

is a well-defined bounded linear functional on $L^1(\mathbb{R})$. Indeed, by the triangle inequality for integrals this defines a functional of norm 1 on $(PC_c(\mathbb{R}), \|\cdot\|_1)$; therefore, it extends uniquely to a continuous functional on the completion $L^1(\mathbb{R})$. Note that since $\|\mathcal{I}\| = 1$, we have for all $f \in L^1(\mathbb{R})$ the inequality $|\mathcal{I}(f)| \le \|f\|_1$, which we may write as

$$\left| \int_{-\infty}^{\infty} f(t)dt \right| \le \int_{-\infty}^{\infty} |f(t)|dt.$$

For brevity, we sometimes write $\int f$ or $\int f(t)dt$ instead of $\int_{-\infty}^{\infty} f(t)dt$.

If $g \in C_b(\mathbb{R})$ or $g \in PC_c(\mathbb{R})$ we can define the multiplication operator

$$M_g : (PC_c(\mathbb{R}), \|\cdot\|_p) \to (PC_c(\mathbb{R}), \|\cdot\|_p), \quad M_g f = gf.$$

The operator $M_g$ is bounded with $\|M_g\| = \|g\|_\infty$ (where $\|g\|_\infty$ is defined to be $\sup\{|g(t)| : t \text{ is a point of continuity}\}$ if $g \in PC_c(\mathbb{R})$), and therefore extends to a bounded operator $M_g : L^p(\mathbb{R}) \to L^p(\mathbb{R})$. This defines a notion of multiplying an $L^p$ function $f$ by a bounded (piecewise) continuous function, which agrees with the familiar notion of multiplication in case that $f$ is actually a function. We summarize this observation by saying that *for every $f \in L^p(\mathbb{R})$ and every $g \in C_b(\mathbb{R}) \cup PC_c(\mathbb{R})$, $gf \in L^p(\mathbb{R})$ and $\|gf\|_p \le \|g\|_\infty \|f\|_p$.*

**Exercise 12.1.6.** Prove that there is a contraction $R : L^p(\mathbb{R}) \to L^p[a, b]$ that extends the restriction operator

$$R : PC_c(\mathbb{R}) \to PC[a, b] \quad , \quad Rf = f\big|_{[a,b]}.$$

$R$ can also be represented as the multiplication operator

$$Rf = f\chi_{[a,b]}.$$

**Exercise 12.1.7.** For $f \in L^1(\mathbb{R})$ and $a, b \in \mathbb{R}$, define

$$\int_a^b f = \int_{-\infty}^\infty f\chi_{[a,b]}.$$

Show that this definition agrees with how one would define $\int_a^b f$ for $f \in L^1[a, b]$. Show that for all $f \in L^1(\mathbb{R})$,

$$\int_{-\infty}^\infty f = \lim_{M,N \to \infty} \int_{-M}^N f.$$

### 12.1.3  *Another construction of the spaces $L^p(\mathbb{R})$

Our construction of $L^p(\mathbb{R})$ has the deficiency, that for many functions — such as $f(x) = e^{-x^2}$ — it is difficult to say whether they do or they do not belong to $L^p(\mathbb{R})$ (recall Remark 2.4.10). We now provide a slightly different definition — one which leads to the same end result, but which makes it easier to identify members of these spaces.

Recall that a function $f : [a, b] \to \mathbb{C}$ is said to be **Riemann integrable on** $[a, b]$, if $f$ is bounded on $[a, b]$, and if there is a number $s \in \mathbb{C}$, such that for every $\epsilon > 0$, there exists a $\delta > 0$ for which the following holds: for every partition $a = t_0 < t_1 < \ldots < t_n = b$ of the interval $[a, b]$ satisfying $\max_i(t_i - t_{i-1}) < \delta$, and every choice of points $\tau_i \in [t_{i-1}, t_i]$,

$$\left| \sum_{i=1}^n f(\tau_i)(t_i - t_{i-1}) - s \right| < \epsilon.$$

In this case, the number $s$ is said to be the **integral of $f$ over** $[a, b]$, and we write $s = \int_a^b f(t)dt$, or simply $s = \int_a^b f$. It is easy to show that if $f$ is Riemann integrable on $[a, b]$, then for every $p \in [1, \infty)$, the function $|f|^p$ is also Riemann integrable on $[a, b]$.

**Definition 12.1.8.** A function $f : [a, b] \to \mathbb{C}$ is said to be **simply GR $p$-integrable** on $[a, b]$ if

1. For all $c, d \in (a, b)$ for which $c < d$, $f$ is Riemann integrable on $[c, d]$.

2. The limit $\lim_{c \to a, d \to b} \int_c^d |f|^p$ exists and is finite.

If $f$ is simply GR $p$-integrable on $[a, b]$, then we define

$$\int_a^b |f|^p = \lim_{c \to a, d \to b} \int_c^d |f|^p.$$

A function $f : [a, b] \to \mathbb{C}$ is said to be **GR $p$-integrable** on $[a, b]$, if there exists a partition $a = t_0 < t_1 < \ldots < t_n = b$ of the interval $[a, b]$ such that $f$ is simply GR $p$-integrable on every interval $[t_{i-1}, t_i]$. In this case we define $\int_a^b |f|^p$ by

$$\int_a^b |f|^p = \sum_{i=1}^n \int_{t_{i-1}}^{t_i} |f|^p.$$

Of course, GR stands for *generalized Riemann*.

**Definition 12.1.9.** Let $GR_0^p(\mathbb{R})$ denote the space of all functions $f : \mathbb{R} \to \mathbb{C}$ such that for every finite interval $[a, b] \subset \mathbb{R}$, the restriction $f|_{[a,b]}$ is GR $p$-integrable on $[a, b]$. For $p \in [1, \infty)$ we denote

$$GR^p(\mathbb{R}) = \left\{ f \in GR_0^p(\mathbb{R}) : \|f\|_p = \sup_{T \geq 0} \left( \int_{-T}^T |f|^p \right)^{1/p} < \infty \right\}.$$

**Proposition 12.1.10.** *Let $\widetilde{L}^p(\mathbb{R})$ denote the completion of $GR^p(\mathbb{R})$ with respect to the norm $\|\cdot\|_p$. Then $\widetilde{L}^p(\mathbb{R})$ can be identified with $L^p(\mathbb{R})$. That is, there exists an isometric isomorphism from $L^p(\mathbb{R})$ onto $\widetilde{L}^p(\mathbb{R})$ that fixes $PC_c(\mathbb{R})$.*

*Proof.* We will show that $PC_c(\mathbb{R})$ is dense in $GR^p(\mathbb{R})$, the rest follows readily (see Exercise 7.5.8).

**Step 1.** *If $f : [a, b] \to \mathbb{C}$ is Riemann integrable, then for every $\epsilon > 0$ there is $g \in PC[a, b]$ such that $\|f - g\|_p < \epsilon$.* Indeed, for $p = 1$ and $f$ real and nonnegative, this follows from the definition of the Riemann integral (take for $g$ the step function $\sum m_i \chi_{[t_{i-1}, t_i]}$, where $m_i = \inf_{t \in [t_{i-1}, t_i]} f(t)$).

If $f : [a, b] \to \mathbb{C}$ and $p = 1$, then we can write $f = f_1 - f_2 + i(f_3 - f_4)$ where $0 \leq f_i \leq |f|$, for $i = 1, 2, 3, 4$. Then we find $g_1, g_2, g_3, g_4 \in PC[a, b]$ such that $\|f_i - g_i\|_1 < \epsilon$, and letting $g = g_1 - g_2 + i(g_3 - g_4)$, we have $\|f - g\|_1 < 4\epsilon$.

Finally, to approximate $f : [a, b] \to \mathbb{C}$ in the norm $\|\cdot\|_p$, we find $g \in PC[a, b]$ such that $\|f - g\|_1 < \epsilon$. Note that if $|f| \leq M$, then the function $g$ we chose also satisfies $|g| \leq M$. We therefore obtain that $\|f - g\|_p^p < M^{p-1}\epsilon$.

**Step 2.** *If $f : [a, b] \to \mathbb{C}$ is simply GR $p$-integrable on $[a, b]$, then for every $\epsilon > 0$, there is some $g \in PC[a, b]$ such that $\|f - g\|_p < \epsilon$.* Since $\lim_{c \to a, d \to b} \int_c^d |f|^p$ exists and is finite, there are $c, d \in (a, b)$ such that $\|f - f\chi_{[c,d]}\|_p < \epsilon/2$. But $f\chi_{[c,d]}$ is Riemann integrable on $[a, b]$, so a function $g \in PC[a, b]$ satisfying $\|g - f\chi_{[c,d]}\|_p < \epsilon/2$ can be found by Step 1.

**Step 3.** *$PC_c(\mathbb{R})$ is dense in $GR^p(\mathbb{R})$.* Given $f \in GR^p(\mathbb{R})$ and $\epsilon > 0$, the assumption $\sup_{T\geq 0}\int_{-T}^{T}|f|^p < \infty$ implies that there is $T > 0$ such that $\|f - f\chi_{[-T,T]}\|_p < \epsilon$. But $f\chi_{[-T,T]}$ is a finite sum of finitely many simply GR $p$-integrable functions. Therefore, by Step 2, $f\chi_{[-T,T]}$ can be approximated as well as we wish by piecewise continuous functions with compact support. $\square$

We can now define $L^p(\mathbb{R})$ to be completion of the normed space $GR^p(\mathbb{R})$ with respect to the norm $\|\cdot\|_p$. Likewise, we can define $L^p[a,b]$ to be the completion of the GR $p$-integrable functions on $[a,b]$ with respect to the norm $\|\cdot\|_p$. The upshot is that if we go by this definition, $L^p(\mathbb{R})$ contains many interesting functions from the outset. For example, if $f(x) = x^{-1/3}$ for $x \in (0,1)$ and $f(x) = 0$ for $x \notin (0,1)$, then it is easy to see that $f \in GR^2(\mathbb{R})$, thus $f \in L^2(\mathbb{R})$. One can also see this way that $f\big|_{[0,1]} \in L^2[0,1]$. Likewise, any function that goes to zero sufficiently fast, such as $e^{-x^2}$, is in $L^p(\mathbb{R})$ for all $p$. In particular, if $f \in C(\mathbb{R})$, then we will say that $f \in L^p(\mathbb{R})$ if $\sup\int_{T}^{T}|f|^p < \infty$, and this happens if and only if $|f|$ is generalized Riemann $p$-integrable on the real line.

---

## 12.2 The Fourier transform on $L^1(\mathbb{R})$

### 12.2.1 Definition and basic properties

**Definition 12.2.1.** Let $f \in L^1(\mathbb{R})$. The ***Fourier transform of*** $f$ is the function $\hat{f}: \mathbb{R} \to \mathbb{C}$ given by

$$\hat{f}(w) = \int_{-\infty}^{\infty} f(t)e^{-iwt}dt. \tag{12.5}$$

**Remark 12.2.2.** Using the notation of the previous section, the Fourier transform is the composition of two bounded operators:

$$\hat{f}(w) = \mathcal{I} \circ M_{e^{-iwt}}(f)$$

for all $f$, where $M_{e^{-iwt}}$ is the multiplication operator $M_{e^{-iwt}}(f)(t) = e^{-iwt}f(t)$ and $\mathcal{I}$ is the integration functional $g \mapsto \int g$.

We let $\mathcal{F}_1$ denote the linear operator given by $\mathcal{F}_1(f) = \hat{f}$. The operator $\mathcal{F}_1$ is also referred to as the ***Fourier transform***. The subscript is there to remind us that this operator is defined on $L^1(\mathbb{R})$.

**Example 12.2.3.** Let $f = \chi_{[a,b]}$. By definition,

$$\hat{f}(w) = \int_{a}^{b} e^{-iwt}dt = \frac{e^{-iwb} - e^{-iwa}}{-iw}$$

for $w \neq 0$, and $\hat{f}(0) = \int_a^b dt = b - a$. We write this as $\mathcal{F}_1(\chi_{[a,b]}) = \frac{e^{-iwa} - e^{-iwb}}{iw}$. In particular,

$$\widehat{\chi_{[-a,a]}}(w) = \frac{2\sin(aw)}{w}.$$

Note that $\mathcal{F}_1(\chi_{[a,b]})$ is continuous and vanishes at $\pm\infty$. Another interesting thing to note is that $\mathcal{F}_1(\chi_{[a,b]})$ is *not* in $L^1(\mathbb{R})$.

Recall that $C_0(\mathbb{R})$ denotes the Banach space of continuous functions vanishing at infinity, which is equipped with the sup norm $\|f\|_\infty = \sup_{t \in \mathbb{R}} |f(t)|$.

**Theorem 12.2.4.** *The operator $\mathcal{F}_1$ given by $\mathcal{F}_1(f) = \hat{f}$ is a bounded linear operator of norm 1 from $L^1(\mathbb{R})$ into $C_0(\mathbb{R})$.*

*Proof.* Let $\mathcal{S}$ be the subspace of $PC_c(\mathbb{R})$ consisting of step functions, i.e., functions of the form

$$\sum_{k=1}^{n} c_k \chi_{[a_k, b_k]}.$$

By Lemma 12.1.4, $\mathcal{S}$ is dense in $L^1(\mathbb{R})$. We will prove that (12.5) defines a bounded linear operator of norm 1 from $\mathcal{S}$ into $C_0(\mathbb{R})$. It then follows by the usual principle of extension of bounded, densely defined operators (Exercise 7.2.9), that this operator extends to a unique bounded operator $\mathcal{F}_1 : L^1(\mathbb{R}) \to C_0(\mathbb{R})$ with the required properties.

Let $f \in \mathcal{S}$. By Example 12.2.3, $\hat{f}$ defined by (12.5) is in $C_0(\mathbb{R})$. If $f \in \mathcal{S}$, then for every $w \in \mathbb{R}$,

$$|\hat{f}(w)| = \left| \int f(t) e^{-iwt} dt \right| \leq \int |f| = \|f\|_1.$$

Moreover, if $f \in \mathcal{S}$ is nonnegative, then

$$\|\hat{f}\|_\infty \geq \hat{f}(0) = \int f(t) dt = \|f\|_1.$$

Thus $f \mapsto \hat{f}$ is a bounded operator of norm 1 from the dense subspace $\mathcal{S} \subset L^1(\mathbb{R})$ into $C_0(\mathbb{R})$, and therefore extends in a unique way to an operator $\mathcal{F}_1 : L^1(\mathbb{R}) \to C_0(\mathbb{R})$ with $\|\mathcal{F}_1\| = 1$. $\square$

**Remark 12.2.5.** The assertion that the Fourier transform $\hat{f}$ of a function $f \in L^1$ satisfies $\hat{f}(w) \to 0$ as $|w| \to \infty$ is often referred to as the *Riemann-Lebesgue lemma*.

For every function $h : \mathbb{R} \to \mathbb{C}$ and every $a \in \mathbb{R}$, we let $h_a$ denote the shifted function $h_a(t) = h(t - a)$. Note that for all $p \in [1, \infty)$, the operation $h \mapsto h_a$ is a linear isometry on $(PC_c(\mathbb{R}), \| \cdot \|_p)$, and therefore extends to a linear isometry of $L^p(\mathbb{R})$.

In many areas of classical analysis, it is customary to abuse notation in any way that makes for better readability, and we shall follow this tradition below. For example, instead of writing $\mathcal{F}_1[f_a]$, we shall write $\mathcal{F}_1[f(t - a)]$. With this in mind, the reader will surely understand what is meant below.

**Proposition 12.2.6.** *Suppose that $f \in L^1(\mathbb{R})$.*

1. *For all $a \in \mathbb{R}$,*

$$\mathcal{F}_1\left[f(t-a)\right] = e^{-iaw}\hat{f}(w) \quad \text{and} \quad \mathcal{F}_1\left[e^{iat}f(t)\right] = \hat{f}(w-a).$$

2. *For all $a > 0$,*

$$\mathcal{F}_1\left[f(at)\right] = \frac{1}{a}\hat{f}(w/a).$$

3. *If $f \in C^1(\mathbb{R}) \cap L^1(\mathbb{R})$ and $f' \in L^1(\mathbb{R})$, then*

$$\mathcal{F}_1\left[f'\right] = iw\hat{f}(w).$$

4. *If $f \in C_c(\mathbb{R})$, then $\hat{f}$ is differentiable and*

$$\mathcal{F}_1\left[tf(t)\right] = i\frac{d}{dw}\hat{f}(w).$$

*Proof.* We leave this as an exercise (the assumptions are carefully stated so that background in Riemann integration will suffice to rigorously prove all items above by easy formal manipulations). $\qquad\square$

**Remark 12.2.7.** Although we shall not need it for the development of the theory, for applications it is useful to know that $\mathcal{F}_1\left[tf(t)\right] = i\frac{d}{dw}\hat{f}(w)$ holds also under weaker assumptions. For example, it is enough to assume that $f \in C_0(\mathbb{R})$ and that $t^2 f(t) \in L^1(\mathbb{R})$ (this will be used in Exercises 12.6.6, 12.6.7 and 12.6.11).

For $k = 1, 2, \ldots, \infty$, we let $C_c^k(\mathbb{R})$ denote the space of $k$-times continuously differentiable functions with compact support on $\mathbb{R}$.

**Corollary 12.2.8.** *If $f \in C_c^2(\mathbb{R})$, then $\hat{f} \in L^1(\mathbb{R})$.*

*Proof.* By Proposition 12.2.6, item (3), we find that $\mathcal{F}_1\left[f''\right] = -w^2\hat{f}(w)$. Being a Fourier transform of an $L^1$ function, we have that $w^2\hat{f}(w) \to 0$ as $|w| \to \infty$. Therefore, the continuous function $\hat{f}$ must be absolutely integrable. $\qquad\square$

### 12.2.2  Convolutions and the convolution theorem

**Definition 12.2.9.** Let $f, g \in PC_c(\mathbb{R})$. The **convolution** of $f$ and $g$, denoted $f * g$, is the function

$$f * g(u) = \int_{-\infty}^{\infty} f(u-t)g(t)dt.$$

**Example 12.2.10.** Let $f = \chi_{[-M,M]}$ and $g = \chi_{[-N,N]}$, and suppose that $M \leq N$. A simple calculation shows that $f * g$ is a continuous piecewise linear function, which is zero up to $-M - N$, then rises linearly until it reaches the value $2M$ at $-N + M$, then continues with this value until $N - M$, and then decreases linearly until reaching 0 at $M + N$, after which it continues as 0.

It is desirable to have the convolution defined for larger classes of functions, not just for the very restricted class $PC_c(\mathbb{R})$. For example, one can prove that $f * g$ is defined if both $f, g$ are in $L^2(\mathbb{R})$, or when $f$ is absolutely integrable and $g$ is bounded. For us, with our minimal requisites in real analysis, it will be most convenient to study the properties of the convolution for pairs of functions in $PC_c(\mathbb{R})$, and later to extend the operation of convolution to larger classes containing $PC_c(\mathbb{R})$ as a dense subspace.

**Proposition 12.2.11.** *For $f, g, h \in PC_c(\mathbb{R})$, we have*

1. $f * g \in C_c(\mathbb{R})$ *(that is $f * g$ is continuous with compact support).*

2. $\|f * g\|_\infty \leq \|f\|_\infty \|g\|_1$ *and* $\|f * g\|_\infty \leq \|f\|_1 \|g\|_\infty$.

3. $\|f * g\|_1 \leq \|f\|_1 \|g\|_1$.

4. $f * g = g * f$.

5. $f * (g + h) = f * g + f * h$ *and* $(f + g) * h = f * h + g * h$.

6. $f * (g * h) = (f * g) * h$.

7. $(f * g)_a = f_a * g = f * g_a$.

*Proof.* Since all functions in sight are piecewise continuous and have compact support, everything can be proved using formal manipulations of the integral, which are justified without any serious mathematics. We leave the last four items as exercises for the reader.

For (2),

$$|f * g(u)| = \left| \int f(u - t)g(t)dt \right| \leq \int |f(u - t)||g(t)|dt \leq \|f\|_\infty \|g\|_1.$$

This gives $\|f * g\|_\infty \leq \|f\|_\infty \|g\|_1$, and the inequality $\|f * g\|_\infty \leq \|f\|_1 \|g\|_\infty$ is proved in a similar manner.

For (3), we exchange order of integration,

$$\int |f * g(u)|du = \int \left| \int f(u - t)g(t)dt \right| du$$
$$\leq \int \int |f(u - t)||g(t)|dtdu$$
$$= \int |g(t)| \left( \int |f(u - t)|du \right) dt$$
$$= \|g\|_1 \|f\|_1.$$

We now prove (1). If the support of $g$ lies in $[a, b]$ and the support of $f$ lies in $[c, d]$, then

$$f * g(u) = \int_a^b f(u - t)g(t)dt.$$

---

The right-hand side of this expression will vanish if $u \notin [-M, M]$ where $M = |a| + |b| + |c| + |d|$, because if $t \in [a,b]$, then $u - t$ has no chance to be in $[-|c| - |d|, |c| + |d|] \supseteq [c,d]$. Thus $f * g$ has compact support.

By Example 12.2.10 and item (5), we see that $f * g \in C_c(\mathbb{R})$ if $f, g \in \mathcal{S}$ (step functions). Now let $f \in PC_c(\mathbb{R})$ and $g \in \mathcal{S}$. If $f_n$ is a sequence of step functions with support contained in the support of $f$ and converging to $f$ in the $L^1$ norm, then the sequence $f_n * g$ consists of continuous functions and by (2) it converges uniformly to $f * g$. Thus, $f * g$ is continuous.

Now let $f, g \in PC_c(\mathbb{R})$. If $g_n$ is a sequence of step functions with support contained in the support of $g$ and converging to $g$ in the $L^1$ norm, then by the previous paragraph, $f * g_n$ is continuous for all $n$, and as above, $f * g_n$ is uniformly convergent to $f * g$. The latter must therefore be continuous, and that concludes the proof of (1). □

We see that the convolution defines a new kind of associative multiplication on $PC_c(\mathbb{R})$. For a fixed $f \in PC_c(\mathbb{R})$, we define the *convolution operator* $C_f$ : $(PC_c(\mathbb{R}), \|\cdot\|_1) \to (PC_c(\mathbb{R}), \|\cdot\|_1)$ by

$$C_f(g) = f * g.$$

By the above proposition, we have that $\|C_f\| \leq \|f\|_1$.

Fixing $f \in PC_c(\mathbb{R})$, since $\|C_f\| \leq \|f\|_1$, we can extend $C_f$ to a contraction on $L^1(\mathbb{R})$, and

$$C_f(g)(u) = \int f(u-t)g(t)dt$$

whenever $g \in L^1(\mathbb{R})$ is such that we know how to define the integral on the right-hand side.

Moreover, given $f \in L^1(\mathbb{R})$, we may also define a convolution operator on $L^1(\mathbb{R})$ as follows. Letting $f_n$ be a sequence in $PC_c(\mathbb{R})$ that converges to $f$ in the $L^1$ norm, we note that items (3) and (5) in Proposition 12.2.11 imply that $C_{f_n}$ is a Cauchy sequence in $B(L^1(\mathbb{R}))$. Let $C_f$ be the limit of this Cauchy sequence. It is easy to show that this definition is independent of the sequence $f_n$ we chose to approximate $f$, and therefore also that this definition of $C_f$ agrees with the previous one when $f$ is in $PC_c(\mathbb{R})$. If one knows measure theory, then one can show that for every $f, g \in L^1(\mathbb{R})$, the operator $C_f$ that we defined acts as

$$C_f(g)(u) = \int f(u-t)g(t)dt$$

for almost every $u$, if we take the integral on the right to be the Lebesgue integral.

If $\phi \in C_0(\mathbb{R})$, then we define the multiplication operator $M_\phi : C_0(\mathbb{R}) \to C_0(\mathbb{R})$ by

$$M_\phi \psi = \phi\psi,$$

where we mean the usual (pointwise) multiplication: $\phi\psi(w) = \phi(w)\psi(w)$ for all $w \in \mathbb{R}$. Note that $\|M_\phi\| = \|\phi\|_\infty$.

**Theorem 12.2.12** (The convolution theorem). *For every $f \in L^1(\mathbb{R})$, the Fourier transform intertwines $C_f$ and $M_{\hat{f}}$, that is*

$$\mathcal{F}_1 C_f = M_{\hat{f}} \mathcal{F}_1. \tag{12.6}$$

*In other words, for every $f, g \in L^1(\mathbb{R})$,*

$$\widehat{f * g} = \hat{f}\hat{g}. \tag{12.7}$$

*Proof.* We will prove (12.7) for $f, g \in PC_c(\mathbb{R})$, because then (12.6) follows by continuity. Indeed, if Equation (12.7) holds for a fixed $f \in PC_c(\mathbb{R})$ and for all $g \in PC_c(\mathbb{R})$, then this shows that (12.6) holds on the dense subspace $(PC_c(\mathbb{R}), \|\cdot\|_1) \subset L^1(\mathbb{R})$; extending continuously we obtain (12.6) for this $f$. Further, if (12.6) holds for all $f \in PC_c(\mathbb{R})$, then whenever $f \in L^1(\mathbb{R})$ we let $\{f_n\}$ be a sequence in $PC_c(\mathbb{R})$ converging to $f$ in the $L^1$ norm, and obtain

$$\|C_f - C_{f_n}\| = \|C_{f-f_n}\| \le \|f - f_n\|_1 \to 0,$$

and

$$\|M_{\hat{f}} - M_{\hat{f}_n}\| = \|M_{\hat{f}-\hat{f}_n}\| = \|\widehat{f - f_n}\|_\infty \le \|f - f_n\|_1 \to 0.$$

Taking the limit in $\mathcal{F}_1 C_{f_n} = M_{\hat{f}_n} \mathcal{F}_1$ we derive (12.6).

It remains to carry out some calculations with piecewise continuous functions. For $f, g \in PC_c(\mathbb{R})$, we are free to make the following manipulations:

$$\begin{aligned}
\widehat{f * g}(w) &= \int e^{-iwu} \int f(u - t)g(t)\,dt\,du \\
&= \int \left( \int e^{-iw(u-t)} f(u - t)\,du \right) e^{-iwt} g(t)\,dt \\
&= \hat{f}(w)\hat{g}(w).
\end{aligned}$$

That concludes the proof. $\qquad\qquad\qquad\qquad\qquad\qquad\qquad\qquad\qquad\square$

## 12.2.3 The pointwise inversion theorem

**Definition 12.2.13.** A function $f : [a, b] \to \mathbb{C}$ is said to be *piecewise $C^1$* if it is piecewise continuous on $[a, b]$, if it is differentiable at all but finitely many points of $[a, b]$, and if $f'$ (which is defined at all but finitely many points of $[a, b]$) is also piecewise continuous. A function $f$ defined on the real line is said to be *piecewise $C^1$* if its restriction to every bounded subinterval is piecewise $C^1$. The space of piecewise $C^1$ and absolutely integrable functions on the real line is denoted by $PC^1(\mathbb{R}) \cap L^1(\mathbb{R})$.

In the above definition, one can interpret *absolutely integrable* in the sense of generalized Riemann integration, that is, a piecewise $C^1$ function $f$ is absolutely integrable if and only if

$$\lim_{M \to \infty} \int_{-M}^{M} |f(t)|\,dt < \infty.$$

**Theorem 12.2.14.** *Let $f \in PC^1(\mathbb{R}) \cap L^1(\mathbb{R})$. For all $t_0 \in \mathbb{R}$,*

$$\frac{f(t_0^+) + f(t_0^-)}{2} = \lim_{R \to \infty} \frac{1}{2\pi} \int_{-R}^{R} \hat{f}(w) e^{iwt_0} dw. \tag{12.8}$$

*In particular, if in addition $f \in C(\mathbb{R})$ and $\hat{f} \in L^1(\mathbb{R})$, then*

$$f(t_0) = \frac{1}{2\pi} \int_{-\infty}^{\infty} \hat{f}(w) e^{iwt_0} dw. \tag{12.9}$$

Note that in (12.8), the integration is carried out over a symmetric interval. It is not claimed that the integral converges in any other sense.

*Proof.* We shall prove (12.8) for $t_0 = 0$, the result for a general point then follows immediately from the $t_0 = 0$ case together Proposition 12.2.6(1).

The right-hand side of (12.8) with $t_0 = 0$ reads

$$\frac{1}{2\pi} \int_{-R}^{R} \hat{f}(w) dw = \frac{1}{2\pi} \int_{-R}^{R} \int_{-\infty}^{\infty} f(t) e^{-iwt} dt dw$$

$$(*) = \frac{1}{2\pi} \int_{-\infty}^{\infty} f(t) \left( \int_{-R}^{R} e^{-iwt} dw \right) dt$$

$$= \frac{1}{2\pi} \int_{-\infty}^{\infty} \frac{e^{iRt} - e^{-iRt}}{it} f(t) dt$$

$$= \frac{1}{\pi} \int_{-\infty}^{\infty} \frac{\sin(Rt)}{t} f(t) dt.$$

Note that the change in the order of integration in $(*)$ is justified, because the function $(t, w) \mapsto e^{-iwt} f(t)$ is absolutely integrable in $\mathbb{R} \times [-R, R]$, and would not be justified if we replaced $R$ with $\infty$. We will show that

$$\lim_{R \to \infty} \frac{1}{\pi} \int_{0}^{\infty} \frac{\sin(Rt)}{t} f(t) dt = \frac{f(0^+)}{2}.$$

In a similar way one shows that the other half of the integral tends to $\frac{f(0^-)}{2}$, and that will conclude the proof. We compute:

$$\frac{1}{\pi} \int_{0}^{\infty} \frac{\sin(Rt)}{t} f(t) dt =$$

$$= \frac{1}{\pi} \int_{0}^{1} \frac{f(t)}{t} \sin(Rt) dt + \frac{1}{\pi} \int_{1}^{\infty} \frac{f(t)}{t} \sin(Rt) dt$$

$$= \frac{1}{\pi} \int_{0}^{1} \frac{f(t)}{t} \sin(Rt) dt + \frac{1}{\pi} \int_{-\infty}^{\infty} g(t) \sin(Rt) dt,$$

where $g(t) = \frac{f(t)}{t}\chi_{[1,\infty]}(t)$. Now $g \in PC^1(\mathbb{R}) \cap L^1(\mathbb{R}) \subset L^1(\mathbb{R})$, and $\sin(Rt) = \frac{e^{iRt}-e^{-iRt}}{2i}$, thus

$$\int g(t)\sin(Rt)dt = \frac{\hat{g}(-R) - \hat{g}(R)}{2i} \xrightarrow{R\to\infty} 0,$$

by Theorem 12.2.4. It remains to show that

$$\lim_{R\to\infty} \frac{1}{\pi}\int_0^1 \frac{f(t)}{t}\sin(Rt)dt = \frac{f(0^+)}{2}.$$

But

$$\frac{1}{\pi}\int_0^1 \frac{f(t)}{t}\sin(Rt)dt =$$
$$= \frac{1}{\pi}\int_0^1 \frac{f(t)-f(0^+)}{t}\sin(Rt)dt + \frac{f(0^+)}{\pi}\int_0^1 \frac{\sin(Rt)}{t}dt$$
$$= \frac{1}{\pi}\int_{-\infty}^{\infty} h(t)\sin(Rt)dt + \frac{f(0^+)}{\pi}\int_0^1 \frac{\sin(Rt)}{t}dt,$$

where $h(t) = \frac{f(t)-f(0^+)}{t}\chi_{[0,1]}(t) \in L^1(\mathbb{R})$. As above we find $\lim_{R\to\infty}\int_{-\infty}^{\infty} h(t)\sin(Rt)dt = 0$.

Finally, using the value of the conditionally convergent integral

$$\int_0^\infty \frac{\sin(u)}{u}du = \pi/2,$$

(see Exercise 4.5.10 for a proof of this identity) we compute

$$\frac{f(0^+)}{\pi}\int_0^1 \frac{\sin(Rt)}{t}dt = \frac{f(0^+)}{\pi}\int_0^R \frac{\sin(u)}{u}du$$
$$\xrightarrow{R\to\infty} \frac{f(0^+)}{2}.$$

That completes the proof. □

## 12.3 The Fourier transform on $L^2(\mathbb{R})$

### 12.3.1 Definition of the Fourier transform on $L^2(\mathbb{R})$

**Lemma 12.3.1.** *If $f \in PC^1(\mathbb{R}) \cap L^1(\mathbb{R})$ and $\hat{f} \in L^1(\mathbb{R})$, then $f, \hat{f} \in L^2(\mathbb{R})$.*

*Proof.* By Theorem 12.2.4, $\hat{f}$ is in $C_0(\mathbb{R})$, and under the assumption $\hat{f} \in L^1(\mathbb{R})$, the inversion formula shows that $f$ is bounded. But a bounded, piecewise continuous function in $L^1(\mathbb{R})$ must be in $L^2(\mathbb{R})$. □

**Theorem 12.3.2** (Plancharel's theorem, I). *For all $f, g \in C_c^2(\mathbb{R})$, we have that $\hat{f}, \hat{g} \in C_0(\mathbb{R}) \cap L^2(\mathbb{R})$, and*

$$2\pi \langle f, g \rangle = \langle \hat{f}, \hat{g} \rangle. \tag{12.10}$$

*In particular,*

$$2\pi \|f\|^2 = \|\hat{f}\|^2. \tag{12.11}$$

*Proof.* By Corollary 12.2.8 and Lemma 12.3.1, we have that $f, g, \hat{f}, \hat{g} \in L^1(\mathbb{R}) \cap L^2(\mathbb{R}) \cap C_0(\mathbb{R})$. Therefore, the following integrals are all absolutely convergent:

$$\langle \hat{f}, \hat{g} \rangle = \int \hat{f}(w)\overline{\hat{g}(w)}dw$$

$$_{(1)} = \int \int f(t)e^{-iwt}\overline{\hat{g}(w)}dtdw$$

$$_{(2)} = \int f(t)\overline{\int \hat{g}(w)e^{iwt}dw}dt$$

$$_{(3)} = 2\pi \int f(t)\overline{g(t)}dt$$

$$= 2\pi \langle f, g \rangle.$$

Since $\hat{f}, \hat{g} \in L^2(\mathbb{R})$, the first integral makes sense. The assumption $f \in L^1(\mathbb{R})$ allows us to consider $\hat{f}$ as the pointwise Fourier transform of $f$ in (1). The fact that $f$ is in $C_c(\mathbb{R})$ and the assumption that $\hat{g} \in L^1(\mathbb{R})$ justifies changing the order of integration in (2). The assumption that $\hat{g} \in L^1(\mathbb{R})$ is used again in (3) for the pointwise inversion of $\hat{g}$. □

Our goal in this section is to define the Fourier transform on $L^2(\mathbb{R})$. Thanks to Plancharel's theorem above, we need only show that $C_c^2(\mathbb{R})$ is dense in $L^2(\mathbb{R})$. We will show a bit more.

**Theorem 12.3.3.** *Let $p \in [1, \infty)$. For all $\epsilon > 0$ and every $f \in PC_c(\mathbb{R})$, there exists $g \in C_c^\infty(\mathbb{R})$ such that*

$$\|f - g\|_p < \epsilon.$$

*Thus, $C_c^\infty(\mathbb{R})$ is dense in $L^p(\mathbb{R})$.*

*Proof.* By Lemma 12.1.4 we only need to show that every step function can be approximated in the $L^p$ norm by $C_c^\infty(\mathbb{R})$ functions. But clearly it is enough for us to be able to approximate characteristic functions of intervals arbitrarily well. Without loss of generality, we will approximate $f = \chi_{[0,1]}$ in the $L^p$ norm using $C_c^\infty(\mathbb{R})$ functions.

We construct an approximating $g$ in several steps as follows.

**Step 1.** Let

$$\phi(t) = \begin{cases} 0 & ,t \leq 0 \\ e^{-1/t^2} & ,t > 0 \end{cases}.$$

**Step 2.** Define

$$\tilde{\psi}(t) = \phi(1 - t^2).$$

Now, $\tilde{\psi}$ is infinitely differentiable, is positive in $(-1, 1)$, and vanishes outside that interval. Let $c = \int \tilde{\psi}$, and define $\psi = c^{-1}\tilde{\psi}$. Then $\psi$ has the same nice properties that $\tilde{\psi}$ enjoyed, with the additional property $\int \psi = 1$.

**Step 3.** For $\epsilon > 0$, define

$$\psi_\epsilon(t) = \epsilon^{-1}\psi(t/\epsilon).$$

Then $\psi_\epsilon$ vanishes off $(-\epsilon, \epsilon)$ and $\int \psi_\epsilon = 1$.

**Step 4.** Now define

$$\theta(t) = \int_{-\infty}^{t} \psi_\epsilon(u) du.$$

$\theta$ is in $C^\infty$, $\theta(t) \in [0, 1]$ for all $t$, $\theta(t) = 0$ for $t < -\epsilon$, and $\theta(t) = 1$ for $t > \epsilon$.

**Step 5.** Finally, we put

$$g(t) = \theta(t) - \theta(t - 1).$$

Then $g \in C_c^\infty(\mathbb{R})$ and

$$\|f - g\|_p^p = \int_{-\infty}^{\infty} |\chi_{[0,1]} - g(t)|^p dt \leq \left( \int_{-\epsilon}^{\epsilon} + \int_{1-\epsilon}^{1+\epsilon} \right) dt = 4\epsilon.$$

$\square$

For an alternative proof, see Theorem 12.5.1 below.

The above theorem, together with the first Plancharel theorem above, allows us to define a Fourier transform on all of $L^2(\mathbb{R})$. The following result, also known as *Plancharel's theorem*, follows immediately by the principle of continuous extension of densely defined operators (Proposition 5.1.6).

**Theorem 12.3.4** (Plancharel's theorem, II). *The Fourier transform extends from $C_c^\infty(\mathbb{R})$ to $L^2(\mathbb{R})$ to give rise to a continuous operator*

$$\mathcal{F}_2 : L^2(\mathbb{R}) \to L^2(\mathbb{R}).$$

*We write $\hat{f} = \mathcal{F}_2(f)$ for $f \in L^2(\mathbb{R})$. $\mathcal{F}_2$ is an isometry-up-to-a-constant, in the sense that*

$$\langle \hat{f}, \hat{g} \rangle = 2\pi \langle f, g \rangle,$$

*for all $f, g \in L^2(\mathbb{R})$. In particular,*

$$\|\hat{f}\|_2^2 = 2\pi \|f\|_2^2,$$

*for all $f \in L^2(\mathbb{R})$.*

Concretely, we may define

$$\hat{f}(w) = L^2 - \lim_{n \to \infty} \frac{1}{2\pi} \int_{-n}^{n} f(t)e^{-iwt}dt.$$

where $L^2 - \lim$ means that $\hat{f}$ is the limit in the $L^2$-norm of the sequence on the right-hand side, that is, $\hat{f} = \lim_n \mathcal{F}_1(f\chi_{[-n,n]})$.

## 12.3.2   The $L^2$-inversion theorem

The operator $\mathcal{F}_2$ is different from $\mathcal{F}_1$, in that $\mathcal{F}_1$ maps $L^1(\mathbb{R})$ into $C_0(\mathbb{R})$, and $\mathcal{F}_2$ maps $L^2(\mathbb{R})$ into $L^2(\mathbb{R})$. In this section, we will see that $\mathcal{F}_2$ is an invertible map. Thus (up to a multiplicative constant) $\mathcal{F}_2$ is a unitary operator.

On $L^2(\mathbb{R})$ we may define a unitary operator, the **reflection operator**, given by

$$\mathcal{R}(f)(t) = f(-t).$$

As usual, we start by defining $\mathcal{R}$ on a dense subspace (say, $C_c(\mathbb{R})$) and then extend it to $L^2(\mathbb{R})$ by continuity.

**Proposition 12.3.5** (Reflection formula). *For all $f \in L^2(\mathbb{R})$,*

$$\hat{\hat{f}}(t) = 2\pi f(-t).$$

*In other words, $\mathcal{F}_2 \circ \mathcal{F}_2 = 2\pi\mathcal{R}$.*

*Proof.* Let $f \in C_c^2(\mathbb{R})$, so that $f, \hat{f} \in L^2(\mathbb{R}) \cap L^1(\mathbb{R}) \cap C_0(\mathbb{R})$. Then, by the inversion formula,

$$\mathcal{F}_2 \circ \mathcal{F}_2(f)(t) = \int \hat{f}(w)e^{-iwt}dw$$
$$= 2\pi\frac{1}{2\pi} \int \hat{f}(w)e^{iw(-t)}dw$$
$$= 2\pi f(-t).$$

Thus, the operators $\mathcal{F}_2 \circ \mathcal{F}_2$ and $2\pi\mathcal{R}$ agree on a dense subspace, and are therefore equal.  □

**Theorem 12.3.6** (The $L^2$-inversion theorem). *The Fourier transform $\mathcal{F}_2 : L^2(\mathbb{R}) \to L^2(\mathbb{R})$ is invertible. In particular, we have the inversion formula*

$$f(t) = L^2 - \lim_{n \to \infty} \frac{1}{2\pi} \int_{-n}^{n} \hat{f}(w)e^{iwt}dw.$$

*Proof.* Since the operator $\mathcal{R}$ is a unitary, the reflection formula shows that $\mathcal{F}_2$ must also be invertible with inverse $\mathcal{F}_2^{-1} = \frac{1}{2\pi}\mathcal{F}_2 \circ \mathcal{R}$; the inversion formula follows.  □

## 12.4  *Shannon's sampling theorem

A function $f : \mathbb{R} \to \mathbb{C}$ is sometimes considered as a continuous signal, where $f(t)$ is interpreted as the value of this signal at time $t$. A natural thing to do from an engineering point of view is to sample a given function at the points of a discrete sequence, that is, to take a series of equally spaced points $\{n\delta\}_{n\in\mathbb{Z}}$ (where $\delta > 0$ is some constant), and extract the sequence of values $\{f(n\delta)\}_{n\in\mathbb{Z}}$.

Considering this process of sampling, one is compelled to ask under what conditions is it possible to reconstruct $f$ from the sequence $\{f(n\delta)\}_{n\in\mathbb{Z}}$. The following theorem gives a precise sufficient condition in terms of the Fourier transform.

**Theorem 12.4.1.** *Let $f : \mathbb{R} \to \mathbb{C}$ be a continuous function in $L^2(\mathbb{R})$. Assume that $\hat{f}(w) = 0$ for all $w \notin [-\frac{\pi}{\delta}, \frac{\pi}{\delta}]$. Then for all $t \in \mathbb{R}$*

$$f(t) = \sum_{n\in\mathbb{Z}} f(n\delta) \frac{\sin(t\pi/\delta - n\pi)}{t\pi/\delta - n\pi}.$$

*Proof.* We can identify $\hat{f}$ with $\hat{f}|_{[-\frac{\pi}{\delta}, \frac{\pi}{\delta}]}$ and consider it as a function in $L^2[-\frac{\pi}{\delta}, \frac{\pi}{\delta}]$. Therefore, $\hat{f}$ is given as the $L^2$ convergent sum

$$\sum_{n\in\mathbb{Z}} c_n e^{in\delta w},$$

where

$$c_n = \frac{\delta}{2\pi} \int_{-\frac{\pi}{\delta}}^{\frac{\pi}{\delta}} \hat{f}(w) e^{-in\delta w} dw.$$

But by the Fourier inversion formula,

$$\frac{\delta}{2\pi} \int_{-\frac{\pi}{\delta}}^{\frac{\pi}{\delta}} \hat{f}(w) e^{-in\delta w} dw = \frac{\delta}{2\pi} \int_{-\infty}^{\infty} \hat{f}(w) e^{-in\delta w} dw = \delta f(-n\delta).$$

We now use the Fourier inversion formula again to obtain

$$f(t) = \frac{1}{2\pi} \int_{-\frac{\pi}{\delta}}^{\frac{\pi}{\delta}} \hat{f}(w) e^{iwt} dw$$

$$= \frac{1}{\delta} \langle \hat{f}(w), e^{-iwt} \rangle_{L^2[-\frac{\pi}{\delta}, \frac{\pi}{\delta}]}$$

$$= \frac{1}{\delta} \left\langle \sum_{n\in\mathbb{Z}} \delta f(-n\delta) e^{in\delta w}, e^{-iwt} \right\rangle_{L^2[-\frac{\pi}{\delta}, \frac{\pi}{\delta}]}$$

$$= \sum_{n\in\mathbb{Z}} f(-n\delta) \frac{\delta}{2\pi} \int_{-\frac{\pi}{\delta}}^{\frac{\pi}{\delta}} e^{i(n\delta+t)w} dw.$$

The value of the last integral is

$$\frac{e^{i(n\delta+t)w}}{i(n\delta+t)}\Big|_{-\pi/\delta}^{\pi/\delta} = \frac{2\sin(n\pi + t\pi/\delta)}{n\delta + t}.$$

Putting everything together we obtain

$$f(t) = \sum_{n\in\mathbb{Z}} f(-n\delta)\frac{\sin(n\pi + t\pi/\delta)}{n\pi + t\pi/\delta} = \sum_{n\in\mathbb{Z}} f(n\delta)\frac{\sin(t\pi/\delta - n\pi)}{t\pi/\delta - n\pi}.$$

□

A function $f$ is said to be **band limited with band $B$** if $\hat{f}$ vanishes outside $[-B, B]$. In other words, $f$ is band limited if it is "made up" from trigonometric exponentials with frequencies only smaller in magnitude than $B$. The sampling theorem says that if one samples a band limited function on a sufficiently dense uniformly spaced sequence, then one loses no information about the original signal. The smaller the band, the larger the spacing we are allowed to leave between two points in the sampling sequence.

**Exercise 12.4.2.** Justify and clarify the use of the Fourier inversion formula in the above proof.

**Exercise 12.4.3.** Prove that

$$H = \{f \in L^2(\mathbb{R}) : \hat{f}(w) = 0 \text{ for all } w \notin [-\pi/\delta, \pi/\delta]\}$$

is a closed subspace of $L^2(\mathbb{R})$, and show that

$$\left\{\frac{\sin(t\pi/\delta - n\pi)}{t\pi/\delta - n\pi}\right\}_{n\in\mathbb{Z}}$$

is an orthonormal basis for $H$.

## 12.5 *The multivariate Fourier transforms

In practice (physics, PDEs, etc.), one often needs to work with spaces of functions in several variables. The spaces $L^p(\mathbb{R}^k)$ can be defined similarly to what we have defined above, as closures of familiar function spaces in the $L^p$ norm

$$\|f\|_p = \left(\int_{\mathbb{R}^k} |f|^p\right)^{1/p}.$$

It is not entirely clear how to generalize the notion of piecewise continuous functions, so we can define $L^p(\mathbb{R})$ to be the closure of the space $C_c(\mathbb{R}^k)$ of continuous functions with compact support, or of the space $\mathcal{S}(\mathbb{R}^k)$ of step functions (those functions which are linear combinations of characteristic functions of boxes $[a_1, b_1] \times \cdots \times [a_k, b_k]$). Any reasonable choice will lead to same space; for definiteness, let us define $L^p(\mathbb{R}^k)$ to be the completion of $C_c(\mathbb{R}^k)$ with respect to the norm $\| \cdot \|_p$. It is quite easy to see that either $\mathcal{S}(\mathbb{R}^k)$ or $C_c(\mathbb{R}^k)$ is dense in $(C_c(\mathbb{R}^k) + \mathcal{S}(\mathbb{R}^k), \| \cdot \|_p)$, therefore both spaces have the same completion (see Exercise 12.6.10).

Given a function $f : \mathbb{R}^k \to \mathbb{C}$ one can define its Fourier transform, defined by

$$\hat{f}(w) = \int_{\mathbb{R}^k} f(x)e^{-iw \cdot x} dx, \tag{12.12}$$

where $x = (x_1, \ldots, x_k)$ and $w = (w_1, \ldots, w_k)$ are $k$-dimensional variables and $w \cdot x = w_1 x_1 + \ldots + w_k x_k$.

Equation (12.12) makes literal sense for all $f \in \mathcal{S}(\mathbb{R}^k) + C_c(\mathbb{R}^k)$, and the operator $f \mapsto \hat{f}$ has properties very similar to one-dimensional transform. In particular, we have $\hat{f} \in C_0(\mathbb{R}^k)$, and $\|\hat{f}\|_\infty \leq \|f\|_1$. Therefore, the map $f \mapsto \hat{f}$ extends to a contraction from $L^1(\mathbb{R}^k)$ into $C_0(\mathbb{R}^k)$, and this operator is called the *(multivariate) Fourier transform*.

Multidimensional convolution is defined in exactly the same way as in the one-dimensional case,

$$f * g(y) = \int_{\mathbb{R}^k} f(y - x)g(x)dx,$$

and the convolution theorem holds: $\widehat{f * g} = \hat{f}\hat{g}$. After one obtains a version of Lemma 12.1.4 that works in several variables, then the same proofs we gave in the one variable case work here as well.

There are versions of the pointwise Fourier inversion theorem, but they are more delicate. However, if $f$ and $\hat{f}$ are both in $C(\mathbb{R}^k) \cap L^1(\mathbb{R}^k)$, then we have

$$f(x) = \frac{1}{(2\pi)^k} \int_{\mathbb{R}^k} \hat{f}(w)e^{iw \cdot x} dw.$$

The development of the Fourier transform on $L^2(\mathbb{R}^k)$ is also carried out in a similar way to the way we developed it above, and similar results hold (with different constants in Plancharel's theorem). Let us concentrate on one thing that goes slightly different.

To extend the Fourier transform to $L^2(\mathbb{R})$, we needed first a dense subspace of functions in which Plancharel's theorem holds. Recall that to carry out the proof of

$$\langle \hat{f}, \hat{g} \rangle = 2\pi \langle f, g \rangle$$

we needed that $f, g, \hat{f}, \hat{g} \in L^2(\mathbb{R}) \cap L^1(\mathbb{R})$. In the one variable case we showed that $C_c^\infty(\mathbb{R}) \subset L^2(\mathbb{R}) \cap L^1(\mathbb{R})$ is dense in both $L^1(\mathbb{R})$ and $L^2(\mathbb{R})$ (and for $f, g \in C_c^\infty(\mathbb{R})$ it is true that $f, g, \hat{f}, \hat{g} \in L^2(\mathbb{R}) \cap L^1(\mathbb{R})$). Recall that we proved this fact by constructing an explicit function that approximates the characteristic function of an interval.

One can prove that $C_c^\infty(\mathbb{R}^k)$ is dense in $L^p(\mathbb{R}^k)$ in a similar fashion to what we did in the one variable case. However, there is a nicer proof that generalizes in many ways, and is worth describing.

**Theorem 12.5.1.** *Let $p \in [1, \infty)$. The space $C_c^\infty(\mathbb{R}^k)$ is dense in $L^p(\mathbb{R}^k)$.*

*Proof.* Let $\phi$ be as in Step 1 of Theorem 12.3.3. Define $\tilde{h}(x) = \phi(1 - \|x\|^2)$, and letting $c = \int \tilde{h}$, put $h = c^{-1}\tilde{h}$. Now for every $\epsilon > 0$ define

$$h_\epsilon(x) = \epsilon^{-k} h(x/\epsilon).$$

The functions $h_\epsilon$ satisfy $\int h_\epsilon = 1$, and also that $h_\epsilon$ vanishes outside the ball $B_\epsilon(0)$ of radius $\epsilon$, and is positive inside that ball.

Up to here, the proof is similar to that of Theorem 12.3.3, but now the proof becomes different. Before moving on, it is worth pointing out that the particular construction of $h$ was not important, just the fact that we had some nonnegative function $h \in C_c^\infty(\mathbb{R}^k)$ which vanishes outside the ball and has integral equal to 1.

Given $f \in C_c(\mathbb{R}^k)$, we define

$$f_\epsilon = f * h_\epsilon.$$

First, note that $f_\epsilon \in C_c^\infty$. Indeed, examining

$$f * h_\epsilon(y) = \int f(y - x) h_\epsilon(x) dx = \int f(x) h_\epsilon(y - x) dx,$$

we see that this function has compact support contained in $\mathrm{supp}(f) + \overline{B_\epsilon(0)}$, and that one may differentiate under the integral sign (with respect to the $y$ variable) infinitely many times.

To see that there is $L^p$ convergence, we compute:

$$|f_\epsilon(y) - f(y)| = \int h_\epsilon(x) |f(y - x) - f(y)| dx$$

$$= \int_{B_\epsilon(0)} h_\epsilon(x) |f(y - x) - f(y)| dx$$

$$\leq \int_{B_\epsilon(0)} h_\epsilon(x) dx \times \max\{|f(y - x) - f(y)| : |x| \leq \epsilon\}$$

$$= \max\{|f(y - x) - f(y)| : |x| \leq \epsilon\},$$

and the last expression tends to 0 as $\epsilon \to 0$ because $f$ is uniformly continuous on $\mathbb{R}^k$. So $f_\epsilon \to f$ uniformly. Since $\mathrm{supp}(f_\epsilon) \subseteq \mathrm{supp}(f) + \overline{B_\epsilon(0)}$, we see that $f_\epsilon \to f$ in the $L^p$ norm as well. This shows that $C_c^\infty(\mathbb{R}^k)$ is dense in $C_c(\mathbb{R}^k)$ in the $L^p$ norm.

The fact that $C_c(\mathbb{R}^k)$ is dense in $L^p(\mathbb{R}^k)$ now implies that $C_c^\infty(\mathbb{R}^k)$ is, too, and the proof is complete. $\square$

## 12.6 Additional exercises

**Exercise 12.6.1.** Let $p \in [1, \infty)$. The map

$$f \mapsto \int_{-\infty}^{\infty} f(t)dt$$

is a well-defined linear functional on $(PC_c(\mathbb{R}), \|\cdot\|_p)$. Prove that this functional is bounded if and only if $p = 1$. If $p \neq 1$, can this linear functional be extended to a linear functional defined on all of $L^p(\mathbb{R})$?

**Exercise 12.6.2.** Compute the Fourier transform of the following functions:

1. $f(x) = e^{-|x|}$.

2. $f(x) = \frac{1}{a^2 + x^2}$.

Use this to

1. Evaluate the integral $\int_0^{\infty} \frac{dx}{a^2 + x^2}$.

2. Find $f$ satisfying the integral equation

$$\int_{-\infty}^{\infty} \frac{f(t)dt}{(x-t)^2 + a^2} = \frac{1}{x^2 + b^2},$$

where $0 < a < b < \infty$.

**Exercise 12.6.3.** Let

$$f(x) = \begin{cases} x & , x \in [-1, 0] \\ 0 & , x \notin [-1, 0] \end{cases}.$$

1. Find $\hat{f}$.

2. Evaluate the integral

$$\int_{-\infty}^{\infty} \frac{1 - \cos t - t \sin t}{2\pi^2(1 + t^2)t^2} dt.$$

**Exercise 12.6.4** (The Poisson summation formula). Let $f \in C^1(\mathbb{R})$, and assume that there exists some constant $C$ such that

$$|f(x)| \leq \frac{C}{1 + |x|^2} \quad \text{and} \quad |f'(x)| \leq \frac{C}{1 + |x|^2},$$

for all $x \in \mathbb{R}$. Prove that

$$\sum_{n \in \mathbb{Z}} f(n) = \sum_{n \in \mathbb{Z}} \hat{f}(2\pi n).$$

(**Hint:** the assumptions imply that $\sum_{n \in \mathbb{Z}} f(x + n)$ converges uniformly to a periodic function.)

**Exercise 12.6.5.** Let $u \in C_c^\infty(\mathbb{R})$, and let $\hat{u}$ be its Fourier transform. For every $n \in \mathbb{N}$, let

$$u_n(x) = \frac{1}{2\pi} \int_{-n}^{n} \hat{u}(w)e^{iwx}dw.$$

Prove that $u_n$ converges uniformly to $u$ on $\mathbb{R}$.

**Exercise 12.6.6.** In this exercise, we will compute the Fourier transform of $f(t) = e^{-t^2}$. In passing, we will also obtain the following definite integral

$$\int_{-\infty}^{\infty} e^{-t^2} dt = \sqrt{\pi}.$$

Since $f \in C^\infty(\mathbb{R}) \cap L^1(\mathbb{R})$, it has a nice Fourier transform $\hat{f}$.

1. Get inspiration from Proposition 12.2.6, Remark 12.2.7, and Corollary 12.2.8 to show that $\hat{f} \in C^1(\mathbb{R})$, that $\hat{f}, \hat{f}' \in L^1(\mathbb{R})$, and that

$$\hat{f}'(w) + \frac{w}{2}\hat{f}(w) = 0.$$

2. Check that the function $e^{-\frac{w^2}{4}}$ is a solution to the above differential equation, and deduce that $\hat{f}(w) = Ce^{-\frac{w^2}{4}}$ where

$$C = \hat{f}(0) = \int_{-\infty}^{\infty} e^{-t^2} dt.$$

3. Use the reflection formula to find that $C = \sqrt{\pi}$. Thus $\hat{f}(w) = \sqrt{\pi}e^{-\frac{w^2}{4}}$.

4. Deduce that

$$\mathcal{F}_2(e^{-\frac{t^2}{2}}) = \widehat{e^{-\frac{t^2}{2}}} = \sqrt{2\pi}e^{-\frac{w^2}{2}}.$$

**Exercise 12.6.7.** Consider the sequence of functions $\{t^n e^{-\frac{t^2}{2}}\}_{n=0}^\infty \subseteq L^2(\mathbb{R})$, and let $\{h_n\}_{n=0}^\infty$ be the orthonormal sequence obtained by applying the Gram-Schmidt process to the former sequence (the functions $h_n$ are called the *Hermite functions*).

1. Prove that for all $n$,
$$h_n(t) = p_n(t)e^{-t^2/2},$$

where $\{p_n\}$ is a sequence of polynomials such that

$$V_n = \text{span}\{p_0 e^{-t^2/2}, \ldots, p_n e^{-t^2/2}\}$$

is equal to $\text{span}\{e^{-t^2/2}, te^{-t^2/2}, \ldots, t^n e^{-t^2/2}\}$ for all $n$.

2. Show that $\mathcal{F}_2(t^k e^{-t^2/2}) = i^k \frac{d^k}{dw^k}\mathcal{F}_2(e^{-t^2/2})$, and conclude that, for all $n$, $V_n$ is a reducing subspace for $\mathcal{F}_2$. Since $\mathcal{F}_2$ is unitary times constant, deduce that $\mathcal{F}_2|_{V_n}$ is diagonalizable.

3. Use the reflection formula to show that

$$\sigma(\mathcal{F}_2) \subseteq \{\sqrt{2\pi}, -\sqrt{2\pi}, i\sqrt{2\pi}, -i\sqrt{2\pi}\}.$$

Deduce that all the eigenvalues of $\mathcal{F}_2|_{V_n}$ have one of the following four values $\sqrt{2\pi}, -\sqrt{2\pi}, i\sqrt{2\pi}, -i\sqrt{2\pi}$.

4. Prove that $\{h_n\}_{n=0}^{\infty}$ is an orthonormal basis for $L^2(\mathbb{R})$. (**Hint:** assuming $f \perp h_n$ for all $n$, it should be shown that $f = 0$. It is helpful to define $g(t) = e^{-t^2/2}f(t)$, and show that $\hat{g} = 0$. Note that with our measure-theory-less definition of $L^2(\mathbb{R})$, some additional explanations are required to justify why this implies that $f = 0$.)

5. Conclude that $\mathcal{F}_2$ is a diagonalizable operator. In the next items of this exercise, which are more challenging computationally, you will find the eigenvectors and eigenvalues.

6. Prove that

$$h_n(t) = C_n e^{t^2/2} \frac{d^n}{dt^n} e^{-t^2} = e^{-\frac{t^2}{2}} H_n(t)$$

where $C_n = \frac{(-1)^n}{(\sqrt{\pi}2^n n!)^{1/2}}$ and $H_n$ is a polynomial of degree $n$ (the polynomials $H_n$ are called the *normalized Hermite polynomials*).

7. Prove that every $h_n$ is an eigenvector, and that the corresponding eigenvalue is $\sqrt{2\pi}i^n$.

**Exercise 12.6.8.** By the previous exercise, $\mathcal{F}_2 : L^2(\mathbb{R}) \to L^2(\mathbb{R})$ is a diagonal unitary (up to a constant). Prove that the operator $\mathcal{F}_1 : L^1(\mathbb{R}) \to C_0(\mathbb{R})$ is injective. Is it surjective? Is it bounded below? Is the range closed? Is $\mathcal{F}_1$ a compact operator? (**Hints:** injectivity is not trivial (yet doable) with our definition of $L^1(\mathbb{R})$. To answer the question about surjectivity, the reader may use the inverse mapping theorem for operators on Banach spaces (a fact that we have not proven in this book). The rest of the questions are easier to answer, but not necessarily in the order they appear).

**Exercise 12.6.9.** True or false: for all $f \in L^1(\mathbb{R})$, the convolution operator $C_f : L^1(\mathbb{R}) \to L^1(\mathbb{R})$ is compact.

**Exercise 12.6.10.** Prove that for all $p \in [1, \infty)$, the space of step functions $\mathcal{S}(\mathbb{R}^k)$ is dense in $L^p(\mathbb{R}^k)$ (in particular, recalling that $L^p(\mathbb{R}^k)$ was defined as the completion of $C_c(\mathbb{R}^k)$, explain why we may consider $\mathcal{S}(\mathbb{R}^k)$ as a subspace of $L^p(\mathbb{R}^k)$).

**Exercise 12.6.11.** The Fourier transform is also useful for the study of certain kinds of differential equations. *The heat equation on an infinite wire* is the partial differential equation with initial conditions

$$u_t(x,t) = cu_{xx}(x,t) \quad , \quad x \in \mathbb{R}, \, t > 0,$$
$$u(x,0) = f(x) \quad \quad , \quad x \in \mathbb{R}.$$

The problem is to find a function $u \in C^2(\mathbb{R} \times [0, \infty))$ that satisfies the conditions above, where $f$ is a given function. We will assume that $f$ is super nice, say $f \in C_c^\infty(\mathbb{R})$. As in Exercise 4.5.17, this problem has a physical interpretation: it describes the evolution of the temperature distribution on an infinite (or very very long) wire.

The problem can be solved as follows. Assuming that $u$ is a solution and assuming also that $u$ is very nice, we "Fourier transform" $u(x, t)$ only with respect to the $x$ variable:

$$\hat{u}(w, t) = \int_{-\infty}^{\infty} u(x, t) e^{-iwx} dx.$$

It can be shown that under some assumptions, $\widehat{u_t} = \frac{\partial}{\partial t} \hat{u}$, so the differential equation becomes

$$\frac{\partial \hat{u}}{\partial t} = c\widehat{u_{xx}} = ciw\widehat{u_x} = -cw^2 \hat{u},$$

because under reasonable assumptions $\mathcal{F}_1(f')(w) = iw\hat{f}(w)$ (recall Proposition 12.2.6). For every $w \in \mathbb{R}$ we can solve this equation, as an ordinary differential equation in the variable $t$. The solution is given by $\hat{u}(w, t) = C_w e^{-cw^2 t}$ (as one may check).

We leave the remainder to you, the reader. Find what $C_w$ must be, and use the inverse transform to recover a formula for the solution $u(x, t)$. Now check directly that this $u$ satisfies the equation and the initial conditions.

# Chapter 13

---

# *The Hahn-Banach theorems

This final chapter is devoted to the Hahn-Banach theorems and some of their consequences. The Hahn-Banach theorems are several closely related results that guarantee the existence of linear functionals with specified properties, and are indispensable in any serious study of functional analysis beyond the setting of Hilbert spaces.

This chapter is the last chapter of this book, but it could just as well be the first chapter in another book on functional analysis.

---

## 13.1 The Hahn-Banach theorems

In this section, we will prove the Hahn-Banach theorem; in fact, we will present five theorems that each can be called "the Hahn-Banach theorem". In the literature, it is common practice to refer to any one of these (or closely related) theorems simply as "the Hahn-Banach theorem"; which particular theorem is used is supposed to be understood from context.

It will be convenient to depart from our convention of working over the complex numbers, and call to the front stage vector spaces over the reals.

### 13.1.1 The Hahn-Banach extension theorems

**Definition 13.1.1.** Let $X$ be a real vector space. A *sublinear functional* is a function $p : X \to \mathbb{R}$ such that

1. $p(x + y) \leq p(x) + p(y)$ for all $x, y \in X$.

2. $p(cx) = cp(x)$ for all $x \in X$ and $c \geq 0$.

**Theorem 13.1.2** (Hahn-Banach extension theorem, sublinear functional version). *Let $X$ be a real vector space, and let $p$ be a sublinear functional on $X$. Suppose that $Y \subseteq X$ is a subspace, and that $f$ is a linear functional on $Y$ such that $f(y) \leq p(y)$ for all $y \in Y$. Then, there exists a linear functional $F$ on $X$ such that $F\big|_Y = f$ and $F(x) \leq p(x)$ for all $x \in X$.*

*Proof.* The first part of the proof involves extending $f$ only to a slightly larger subspace. We will later use the first part to show that $f$ can be extended all the way up to $X$.

Let $x \notin Y$ be nonzero, and define

$$W = \{y + cx : y \in Y, c \in \mathbb{R}\}.$$

The goal is to extend $f$ to a functional $F$ on $W$ such that $F(w) \leq p(w)$ for all $w \in W$. Since $x$ is independent of $Y$, we are free to define $F(x)$ to be any real number, and that determines uniquely a linear extension $F$ of $f$, given by $F(y + cx) = f(y) + cF(x)$. The issue here is to choose $F(x)$ so that $F$ is smaller than $p$ on $W$. Now having chosen the value $F(x)$, the requirement $F(w) \leq p(w)$ for all $w \in W$ is equivalent to

$$f(y) + cF(x) = F(y) + cF(x) = F(y + cx) \leq p(y + cx) \qquad (13.1)$$

for all $y \in Y, c \in \mathbb{R}$. If $c = 0$, then (13.1) becomes $f(y) \leq p(y)$, which is satisfied by assumption. If $c \neq 0$, we may divide (13.1) by $|c|$, and we may replace $y$ by a multiple of $y$, to convert condition (13.1) to

$$F(x) \leq p(x - y) + f(y) \text{ and } f(y) - p(y - x) \leq F(x) \qquad (13.2)$$

for all $y \in Y$. To summarize, if we set the value $F(x)$ and define $F$ on $W$ by $F(y + cx) = f(y) + cF(x)$, then (13.1) holds for all $y \in Y$ and all $c \in \mathbb{R}$, if and only if (13.2) holds for all $y \in Y$.

Therefore, if there exists any $t \in \mathbb{R}$ such that

$$\sup\{f(z) - p(z - x) : z \in Y\} \leq t \leq \inf\{p(x - y) + f(y) : y \in Y\}, \qquad (13.3)$$

then we can define $F(x) = t$, and that determines a linear extension $F$ defined on $W$, that is dominated by $p$. To see that there exists $t$ fulfilling (13.3), it suffices to show that the supremum appearing there is smaller than the infimum. But for all $y, z \in Y$,

$$f(z) - f(y) = f(z - y) \leq p(z - y) \leq p(z - x) + p(x - y),$$

or $f(z) - p(z - x) \leq f(y) + p(x - y)$. We conclude that a $t$ satisfying (13.3) exists, and the first part of the proof is complete.

And now for the second part of the proof. Let $\mathcal{P}$ be the collection of all pairs $(g, Z)$ such that

1. $Z$ is a linear subspace of $X$ containing $Y$.

2. $g$ is a linear functional on $Z$ that extends $f$.

3. $g(z) \leq p(z)$ for all $z \in Z$.

The pair $(Y, f)$ is in $\mathcal{P}$, so $\mathcal{P}$ is not empty. Let $\mathcal{P}$ be partially ordered by the rule $(Z, g) \leq (Z', g')$ if and only if $Z' \supseteq Z$ and $g'|_Z = g$. It is easy to see that every chain in $\mathcal{P}$ has an upper bound, thus by Zorn's lemma $\mathcal{P}$ has a maximal element $(\hat{Z}, \hat{g})$. Now, $\hat{Z}$ must be $X$, otherwise, there would exist $x \notin \hat{Z}$, and then, by the first part of the proof, we would be able to extend $\hat{g}$ to the space $\{z + cx : z \in \hat{Z}, c \in \mathbb{R}\}$. But that would contradict maximality, thus we conclude that $\hat{Z} = X$, and $F = \hat{g}$ is the required extension of $f$. $\quad\square$

Recall that for any normed space $X$, the dual space $X^*$ is defined to be the space of all bounded linear functionals from $X$ into the scalar field, with norm given by

$$\|F\| = \sup_{x \in X_1} |F(x)| \quad , \quad F \in X^*,$$

where $X_1 = \{x \in X : \|x\| \leq 1\}$.

**Theorem 13.1.3** (Hahn-Banach extension theorem, bounded functional version). *Let $Y$ be a subspace of a normed space $X$, and let $f \in Y^*$. Then, there exists $F \in X^*$ such that $F|_Y = f$ and $\|F\| = \|f\|$.*

*Proof.* We prove the theorem first for the case where $X$ is a space over the reals. Define $p(x) = \|f\|\|x\|$ for all $x \in X$. This is easily seen to be a sublinear functional on $X$, which dominates $f$ on $Y$. By Theorem 13.1.2, there exists $F$ extending $f$ such that $|F(x)| \leq p(x) = \|f\|\|x\|$ for all $x$, thus $\|F\| \leq \|f\|$. Since $F$ extends $f$, it follows that $\|F\| = \|f\|$.

Suppose now that $X$ is a normed space over the complex numbers. Then $X$ and $Y$ are also normed spaces over the reals, and $Y$ is a subspace of $X$ also when these are considered as real linear spaces. Define $g = \operatorname{Re} f$. Then $g$ is a bounded real functional on $Y$, and $\|g\| \leq \|f\|$. By the previous paragraph, $g$ extends to a bounded real functional $G$ on $X$ such that $\|G\| = \|g\|$.

We now define $F(x) = G(x) - iG(ix)$. Some computations show that $F$ is linear and extends $f$. To get $\|F\| = \|f\|$, it suffices to show that $\|F\| \leq \|G\|$.

Fix $x \in X$, and write $F(x) = re^{it}$. Then $|F(x)| = r = e^{-it}F(x) = F(e^{-it}x) = G(e^{-it}x) - iG(ie^{-it}x)$. But this is a real number, so its imaginary part vanishes and we get $|F(x)| = G(e^{-it}x) \leq \|G\|\|x\|$, which proves $\|F\| \leq \|G\|$. $\quad\square$

**Exercise 13.1.4.** Let $X$ be a vector space over the complex numbers. $X$ can also be considered as a real vector space. Show that for every complex linear functional $F : X \to \mathbb{C}$, the real part of $F$ (the functional $x \mapsto \operatorname{Re} F(x)$) is a real linear functional on the real space $X$, and that

$$F = \operatorname{Re} F(x) - i \operatorname{Re} F(ix), \quad x \in X.$$

Deduce that $F \mapsto \operatorname{Re} F$ is a bijection between complex linear functionals and real linear functionals on $X$.

**Exercise 13.1.5.** Show that in the above proof, the functional $F$ is a linear functional that extends $f$.

## 13.1.2   The Hahn-Banach separation theorems

**Theorem 13.1.6** (Hahn-Banach separation theorem, subspace/point). *Let M be a linear subspace of a normed space $X$, and let $x \in X$. Put*

$$d = d(x, M) = \inf\{\|x - m\| : m \in M\}.$$

*Then, there exists $F \in X_1^* = \{G \in X^* : \|G\| \le 1\}$, such that $F(x) = d$ and $F|_M = 0$.*

*Proof.* On span$\{x, M\}$ we define a linear functional by $f(cx + m) = cd$. For every $c \ne 0$ and every $m \in M$,

$$\|cx + m\| = |c|\|x - (-c^{-1}m)\| \ge |c|d,$$

thus

$$|f(cx + m)| = |cd| \le \|cx + m\|.$$

This shows that $\|f\| \le 1$. The Hahn-Banach extension theorem provides $F \in X^*$ which satisfies all the requirements. $\qquad\square$

**Corollary 13.1.7.** *Let $X$ be a normed space and $x \in X$. Then, there exists $F \in X_1^*$ for which $F(x) = \|x\|$.*

*Proof.* This follows Theorem 13.1.6 applied to $M = \{0\}$. $\qquad\square$

**Corollary 13.1.8.** *Let $x$ be an element in a normed space $X$ for which $F(x) = 0$ for all $F \in X^*$. Then $x = 0$.*

**Corollary 13.1.9.** *Let $M$ be a subspace in a normed space. A point $x \in X$ is in $\overline{M}$ if and only if $F(x) = 0$ for all $F \in X^*$ that vanishes on $M$.*

*Proof.* One implication follows from continuity of $F$. For the other implication, we use Theorem 13.1.6 to see that if $F(x) = 0$ for all $F$ that vanishes on $M$, then $d(x, M) = 0$, in other words $x \in \overline{M}$. $\qquad\square$

Let us fix for the rest of the chapter a normed space $X$ over the reals. Since every complex space is also a real space, the following separation theorems can be applied to complex spaces (the statements and definitions should be modified by replacing every functional by its real part).

Theorem 13.1.6 says that there is a functional $F \in X_1^*$ that is a witness to the fact that the distance of $x$ from $M$ is $d$. This is sometimes restated by saying that $F$ "separates" $M$ from $x$. To make the notion of separation more geometrically intuitive, we introduce the following terminology.

**Definition 13.1.10.** A *hyperplane* in $X$ is a subset of the form

$$F^{-1}(c) = \{x \in X : F(x) = c\},$$

where $F$ is a nonzero linear functional on $X$ and $c \in \mathbb{R}$.

**Definition 13.1.11.** If $A, B \subseteq X$, we say that the hyperplane $F^{-1}(c)$ *separates* $A$ from $B$ if

$$F(A) \leq c \leq F(B).$$

We say that this hyperplane *strictly separates* $A$ from $B$ if there is some $\epsilon > 0$ such that

$$F(A) \leq c - \epsilon < c + \epsilon \leq F(B).$$

It is instructive to draw a picture that goes with this definition. Draw two convex bodies $A$ and $B$ in the plane, so that their boundaries are disjoint. No matter how you drew $A$ and $B$, so long as the bodies are convex and do not touch, you will be able to draw a straight line that cuts the plane into two halves, one containing $A$ and the other containing $B$. A straight line in the plane is given by an equation of the form $ax + by = c$, so we can say, in the language we just introduced, that whenever we are given two disjoint convex bodies $A$ and $B$ in the plane, we can find a hyperplane that separates $A$ from $B$. The Hahn-Banach separation theorems that we will prove below give a precise formulation of this separation phenomenon in the setting of any real normed space $X$.

In infinite dimensional spaces, a hyperplane is not necessarily a nice set. To make our geometric vocabulary complete, we note that if $F$ is bounded, then $F^{-1}(c)$ is closed. The converse is also true.

**Exercise 13.1.12.** Prove that a hyperplane $F^{-1}(c)$ is closed if and only if it is not dense, and this happens if and only if $F$ is bounded. (**Hint:** treat first the case $c = 0$; recall that part of the definition of hyperplane is that $F \neq 0$.)

To obtain the separation theorems, we will make use of the following device.

**Definition 13.1.13** (The Minkowski functional). Let $C \subseteq X$ be a convex and open set. Define $p : X \to [0, \infty)$ by

$$p(x) = \inf\{t > 0 : t^{-1}x \in C\}.$$

$p$ is called the *Minkowski functional* of $C$.

**Lemma 13.1.14.** *Let $C \subseteq X$ be a convex and open set containing $0$. Then, $p$ is a sublinear functional, $C = \{x \in X : p(x) < 1\}$, and there exists some $M > 0$ such that $p(x) \leq M\|x\|$ for all $x \in X$.*

*Proof.* There is some $r$ such that the closed ball $B(0, r) \subset C$, thus for every $x \neq 0$, $\frac{r}{\|x\|}x \in C$, hence $p(x) \leq \frac{1}{r}\|x\|$. That takes care of the last assertion.

Assume that $x \in C$. The set $C$ is open, so $(1 + r)x \in C$ for sufficiently small $r > 0$, whence $p(x) \leq (1 + r)^{-1} < 1$.

Assume that $p(x) < 1$. Thus there is some $0 < t < 1$ for which $t^{-1}x \in C$. But since $x$ is on the ray connecting $0$ and $t^{-1}x$ and $C$ is convex, $x \in C$. We conclude that $C = \{x \in C : p(x) < 1\}$.

Clearly, $p(0) = 0$. For $c > 0$, $p(cx)$ can be written as

$$\inf\{ct > 0 : (ct)^{-1}cx \in C\} = c\inf\{t > 0 : t^{-1}x \in C\} = cp(x).$$

We proceed to prove that $p$ is subadditive. Let $x, y \in X$ and $r > 0$. From the definition $(p(x)+r)^{-1}x$ and $(p(y)+r)^{-1}y$ are in $C$. Every convex combination

$$t(p(x) + r)^{-1}x + (1 - t)(p(y) + r)^{-1}y , \quad t \in [0, t] \tag{13.4}$$

is also in $C$. We now choose the value of $t$ cleverly, so that the coefficients of $x$ and $y$ will be the same, and in that way we will get something times $x + y$. The solution of the equation

$$t(p(x) + r)^{-1} = (1 - t)(p(y) + r)^{-1}$$

is $t = \frac{p(x)+r}{p(x)+p(y)+2r}$, and plugging that value of $t$ in (13.4) gives $(p(x) + p(y) + 2r)^{-1}(x + y) \in C$. Thus, $p(x + y) \le p(x) + p(y) + 2r$. Letting $r$ tend to 0 we obtain the result. $\qquad\square$

**Theorem 13.1.15** (Hahn-Banach separation theorem, convex/open). *Let $A, B \subseteq X$ be two nonempty disjoint convex sets, and suppose that $B$ is open. Then there exists a closed hyperplane which separates $A$ and $B$.*

*Proof.* Let us first treat the case where $0 \in B$ and $A$ consists of one point, say $A = \{a\}$. Let $p$ be the Minkowski functional of $B$. Define a linear functional $f : \text{span}\{a\} \to \mathbb{R}$ by

$$f : \lambda a \mapsto \lambda.$$

By Lemma 13.1.14, $p(a) \ge 1$. Therefore if $\lambda \ge 0$, then $f(\lambda a) = \lambda \le \lambda p(a) = p(\lambda a)$. If $\lambda < 0$, then $f(\lambda a) < 0 \le p(\lambda a)$. We see that $f(\lambda a) \le p(\lambda a)$ for all $\lambda$.

By Theorem 13.1.2, $f$ can be extended to a functional $F$ on $X$ such that $F(x) \le p(x)$ for all $x \in X$. By the last part of Lemma 13.1.14, $F$ is bounded. This $F$ satisfies $F(a) = 1$ and $F(x) \le p(x) < 1$ for all $x \in B$, so the closed hyperplane $F^{-1}(1)$ separates $A$ and $B$, and in fact

$$F(b) < 1 = F(a) \quad \text{for all } b \in B.$$

If $A = \{a\}$ and $B$ is open but does not contain 0, then we choose some $b_0 \in B$ and apply the result from the previous paragraph to the sets $\{a - b_0\}$ and $B - b_0 = \{b - b_0 : b \in B\}$.

Now let us treat the general case. The set $A - B = \{a - b : a \in A, b \in B\}$ is convex and open, and is disjoint from the set $\{0\}$. By the previous paragraph, there is a closed hyperplane that separates $A - B$ and $\{0\}$ as follows

$$F(a - b) < F(0) = 0 \quad \text{for all } a \in A, b \in B,$$

so $F(a) < F(b)$ for all $a, b$. If $c$ satisfies $\sup F(A) \le c \le \inf F(B)$, then $F^{-1}(c)$ separates $A$ and $B$. $\qquad\square$

**Exercise 13.1.16.** Prove that if $A$ is convex and if $B$ is convex and open, then the set $A - B$ is convex and open.

**Theorem 13.1.17** (Hahn-Banach separation theorem, compact/closed). *Let $A, B \subseteq X$ be two nonempty disjoint convex sets, such that $A$ is closed and $B$ is compact. Then there exists a closed hyperplane which strictly separates $A$ and $B$.*

*Proof.* As above, we consider $A - B$, call this set $C$. As above, $C$ is convex and does not contain 0. Since $B$ is compact, it follows that $C$ is closed. Let $B(0,r)$ be a small ball disjoint from $C$. By Theorem 13.1.15, there is a functional $F \in X^*$ and $c \in \mathbb{R}$ such that

$$F(a) - F(b) = F(a - b) \leq c \leq \inf F(B(0,r)) = -\|F\|r$$

for all $a \in A, b \in B$. It follows that $A$ and $B$ can be strictly separated by some hyperplane defined by $F$. $\qquad\square$

**Exercise 13.1.18.** Prove that if $A$ is closed and $B$ is compact, then the set $A - B$ is closed. What happens if $B$ is not assumed to be compact?

### 13.1.3   Banach limit

The Hahn-Banach theorem has countless applications in analysis. The purpose of this section is to present an application to sequence spaces, which does not require any knowledge in function theory or measure theory. The reader is referred to [18, Chapter 9] for several serious applications.

Let $\ell^\infty$ be the Banach space over the reals, consisting of all real bounded sequences, equipped with the supremum norm

$$\ell^\infty = \left\{ a = (a_n)_{n=0}^\infty : \|a\|_\infty = \sup_{n \geq 0} |a_n| < \infty \right\}.$$

Note that we have previously used the same notation for the space of bounded sequences of complex numbers; this minor overloading of notation should not cause any harm. The space $\ell^\infty$ contains a closed subspace

$$c = \left\{ a \in \ell^\infty : \lim_{n \to \infty} a_n \text{ exists} \right\}.$$

The functional $f : a \mapsto \lim_n a_n$ is a bounded linear functional on $c$. By the Hahn-Banach theorem, this functional can be extended to a bounded linear functional $F$ defined on all of $\ell^\infty$. We will show that $F$ can be chosen to have some of the nice properties that $f$ has. For example, if we let $S$ denote the backward shift operator

$$S((a_0, a_1, a_2, \ldots)) = (a_1, a_2, \ldots),$$

then $f$ is translation invariant, in the sense that $f(Sa) = f(a)$ for all $a \in c$.

**Theorem 13.1.19** (Banach limit). *There exists a bounded linear functional $L$ on $\ell^\infty$, such that for all $a = (a_n)_{n=0}^\infty$,*

1. *$L(Sa) = L(a)$,*

2. *If $m \le a_n \le M$ for all $n$, then $m \le L(a) \le M$,*

3. *If $\lim_n a_n$ exists, then $L(a) = \lim_n a_n$.*

**Remark 13.1.20.** A functional on $\ell^\infty$ satisfying the above conditions is called a **Banach limit**.

*Proof.* Let $M$ be the range of the operator $I - S$, that is

$$M = \{a - Sa : a \in \ell^\infty\}.$$

$M$ is a subspace, and every $b \in M$ has the form

$$b = (b_n)_{n=0}^\infty = (a_1 - a_0, a_2 - a_1, \ldots).$$

Such an element $b$ satisfies $\sum_{k=0}^n b_k = a_{n+1} - a_0$. In particular

$$\sup_n \left| \sum_{k=0}^n b_k \right| < \infty. \tag{13.5}$$

If we let $\mathbf{1} = (1, 1, 1, \ldots)$, then $\mathbf{1} \notin M$. We claim that

$$d(\mathbf{1}, M) = \inf\{\|\mathbf{1} - b\| : b \in M\} = 1.$$

Indeed, if there was a $b \in M$ such that $\|\mathbf{1} - b\| < 1$, this would contradict (13.5). By Theorem 13.1.6, there exists $L \in (\ell^\infty)^*$ such that $\|L\| = 1$, $L(\mathbf{1}) = 1$ and $L\big|_M = 0$. Since $L(b) = 0$ for all $b \in M$, we see that $L(Sa) = L(a)$ for all $a \in \ell^\infty$.

Next, suppose that $a_n \ge 0$ for all $n \in \mathbb{N}$. Then, consider the sequence

$$x = \|a\|\mathbf{1} - a = (\|a\| - a_0, \|a\| - a_1, \|a\| - a_2, \ldots).$$

Since $\|x\| \le \|a\|$ and $\|L\| = 1$, we get

$$\|a\| - L(a) = L(x) \le \|L\|\|x\| \le \|a\|,$$

thus $L(a) \ge 0$. From here, it is easy to prove that $L$ satisfies the second property.

Finally, to prove that $L(a) = \lim_n a_n$ when the limit exists, we will show that for any $a \in \ell^\infty$,

$$\liminf_n a_n \le L(a) \le \limsup_n a_n.$$

We will show the second inequality; the first one is shown in a similar manner.

For this, consider the sequence $y = (a_k, a_{k+1}, a_{k+2}, \ldots)$. Then $y_n \leq \sup_{m \geq k} a_m$ for all $n$, and therefore

$$L(a) = L(y) \leq \sup_{m \geq k} a_m,$$

whence $L(a) \leq \limsup_n a_n$.      □

We can now finally settle a question that we raised in Chapter 7. Recall that in Example 7.3.5 we noted that there exists an isometric map

$$\ell^1 \ni b \mapsto \Gamma_b \in (\ell^\infty)^*,$$

where $\Gamma_b$ is given by

$$\Gamma_b(a) = \sum a_n b_n.$$

Motivated by the case where the roles of $\ell^1$ and $\ell^\infty$ are reversed and by the case of $\ell^p$ spaces where $p \neq 1, \infty$ (see Example 7.3.4 and Exercise 7.3.6), one is naturally led to ask whether the map $b \mapsto \Gamma_b$ is surjective.

**Exercise 13.1.21.** Show that the map $b \mapsto \Gamma_b$ discussed above is not surjective.

Note that the above exercise only settles the question whether a particular map is an isometric isomorphism between $\ell^1$ and $(\ell^\infty)^*$, that is, the exercise shows that the map given by $b \mapsto \Gamma_b$ is not an isometric isomorphism. But it leaves open the question whether $\ell^1$ is isometrically isomorphic (or maybe just isomorphic) to the dual of $\ell^\infty$. We will answer this question in the next section.

---

## 13.2   The dual space, the double dual, and duality

In Chapter 7 we defined the dual space $X^*$ of a Banach space $X$, but we did not prove anything significant regarding dual spaces, because we had no tools to deal with them. For example, we could not assert that the limit of a weakly convergent sequence is unique, nor could we determine whether or not $(\ell^\infty)^* \cong \ell^1$. In fact, we could not even say that a general normed space has any nonzero bounded functional defined on it.

Now with the Hahn-Banach theorems at our hands, we know that every normed space has lots of bounded functionals on it — enough to separate points. Corollary 13.1.8 is a generalization of the fact that in an inner product space $G$, if $\langle g, h \rangle = 0$ for all $h \in G$, then $g = 0$. Corollary 13.1.9 is a generalization of the fact that in $\overline{M} = M^{\perp\perp}$ (Corollary 3.2.16).

One can learn many things about a Banach space $X$ from its dual $X^*$. For example:

**Exercise 13.2.1.** Prove that if $X^*$ is separable, then $X$ is separable, too.

Now we can finally answer the question raised in Chapter 7 and in the previous section.

**Exercise 13.2.2.** Prove that $(\ell^\infty)^*$ is not isomorphic to $\ell^1$.

**Definition 13.2.3.** Let $X$ be a normed space. Then $X^*$ is a Banach space in its own right, and therefore has a dual space

$$X^{**} = (X^*)^*.$$

The space $X^{**}$ is called the **double dual** (or **bidual**) of $X$.

Every $x \in X$ gives rise to a function $\hat{x} \in X^{**} = (X^*)^*$ by way of

$$\hat{x}(f) = f(x), \quad f \in X^*.$$

**Proposition 13.2.4.** *The map $x \mapsto \hat{x}$ is an isometry from $X$ into $X^{**}$.*

*Proof.* Linearity is trivial. To see that the map is norm preserving,

$$\|\hat{x}\| = \sup_{f \in X_1^*} |f(x)| = \|x\|,$$

where the first equality is true by definition of the norm of a functional, and the second equality follows from Corollary 13.1.7. □

**Definition 13.2.5.** A Banach space $X$ is said to be **reflexive** if the map $x \mapsto \hat{x}$ is an isometry of $X$ onto $X^{**}$.

**Example 13.2.6.** Every Hilbert space is reflexive.

**Example 13.2.7.** The spaces $\ell^p$ are reflexive for all $p$ in the range $(1, \infty)$ (this follows from Exercise 7.3.6). It is also true that the spaces $L^p[0,1]$ are reflexive for all $p \in (1, \infty)$.

**Example 13.2.8.** The space $\ell^1$ is an example of a nonreflexive space (this follows from Exercise 13.1.21).

It is sometimes convenient, when considering the action of $f \in X^*$ on $x \in X$, to write

$$f(x) = \langle f, x \rangle.$$

This is just a matter of notation, and invites one to think of the relationship between $X$ and $X^*$ as something more symmetrical. Elements of the dual space $X^*$ are sometimes denoted by starred symbols, for example $x^*$. Thus we have

$$x^*(x) = \langle x^*, x \rangle = \hat{x}(x^*) = \langle \hat{x}, x^* \rangle, \quad x \in X, x^* \in X^*.$$

Since $X$ norms $X^*$ (i.e., $\|x^*\| = \sup_{x \in X_1} |\langle x^*, x \rangle|$) and $X^*$ norms $X$ (i.e., $\|x\| = \sup_{x^* \in X_1^*} |\langle x^*, x \rangle|$), there is some symmetry between $X$ and its dual $X^*$. The two spaces are said to be *in duality*. One should keep in mind, though, that the situation is not entirely symmetrical when $X$ is not reflexive.

**Definition 13.2.9.** Let $X$ be a normed space and let $S \subseteq X$. The **annihilator of** $S$, denoted $S^\perp$, is the set

$$S^\perp = \{x^* \in X^* : \langle x^*, x \rangle = 0 \text{ for all } x \in S\}.$$

**Definition 13.2.10.** Let $X$ be a normed space and let $T \subseteq X^*$. The **pre-annihilator of** $T$, denoted $^\perp T$, is the set

$$^\perp T = \{x \in X : \langle x^*, x \rangle = 0 \text{ for all } x^* \in T\}.$$

With this notation we can state Corollary 13.1.9 as follows: *A subspace $M$ contains $x$ in its closure if and only if $x \in^\perp (M^\perp)$.* In other words,

$$\overline{M} = {}^\perp(M^\perp). \tag{13.6}$$

**Definition 13.2.11.** Let $T \in B(X, Y)$. The **adjoint** of $T$ is the operator $T^* : Y^* \to X^*$ given by

$$T^* y^* = y^* \circ T.$$

**Theorem 13.2.12.** *Let $X$ and $Y$ be normed spaces, and let $T \in B(X, Y)$. Then $T^* \in B(Y^*, X^*)$, and $\|T^*\| = \|T\|$. For all $x \in X, y^* \in Y^*$,*

$$\langle y^*, Tx \rangle = \langle T^* y^*, x \rangle.$$

*Moreover, $T^*$ is the unique map from $Y^*$ to $X^*$ with this property.*

**Exercise 13.2.13.** Prove Theorem 13.2.12.

**Theorem 13.2.14.** *Let $X$ and $Y$ be normed spaces, and let $T \in B(X, Y)$. Then $\ker(T^*) = \operatorname{Im}(T)^\perp$ and $\ker(T) = {}^\perp \operatorname{Im}(T^*)$.*

*Proof.* Thanks to the notation, the proof is exactly the same as in the Hilbert space case (note, though, that both assertions require proof, and do not follow one from the other by conjugation). $\qquad \square$

**Corollary 13.2.15.** *For $T \in B(X, Y)$, $^\perp \ker(T^*) = \overline{\operatorname{Im}(T)}$.*

*Proof.* This follows from the theorem and (13.6). $\qquad \square$

---

## 13.3  Quotient spaces

Let $X$ be a normed space, and let $M \subseteq X$ be a closed subspace. In linear algebra, one learns how to form the quotient space $X/M$, which is defined to be the set of all cosets $x + M$, $x \in X$, with the operations

$$(x + M) + (y + M) = x + y + M.$$

$$c(x + M) = cx + M.$$

It is known and also easy to show that these operations are well-defined, and give $X/M$ the structure of a vector space. Our goal now is to make $X/M$ into a normed space, and to prove that when $X$ is complete, so is $M$.

To abbreviate, let us write $\dot{x}$ for $x + M$. We define

$$\|\dot{x}\| = d(x, M) = \inf\{\|x - m\| : m \in M\}. \tag{13.7}$$

Let $\pi$ denote the quotient map $X \to X/M$.

**Theorem 13.3.1.** *Let $X$ be a normed space, $M \subseteq X$ a closed subspace, and define the quotient space $X/M$ as above. With the norm defined as in (13.7), the following hold:*

1. *$X/M$ is a normed space.*

2. *$\pi$ is a contraction: $\|\pi(x)\| \leq \|x\|$ for all $x$.*

3. *For every $y \in X/M$ such that $\|y\| < 1$, there is an $x \in X$ with $\|x\| < 1$ such that $\pi(x) = y$.*

4. *$U$ is open in $X/M$ if and only if $\pi^{-1}(U)$ is open in $X$.*

5. *If $F$ is a closed subspace containing $M$, then $\pi(F)$ is closed.*

6. *If $X$ is complete, then so is $X/M$.*

*Proof.* For every $x \in X$ and every scalar $c$, the identity $\|c\dot{x}\| = |c|\|\dot{x}\|$ is trivial. Next, $\|\dot{x}\| = 0 \Leftrightarrow \dot{x} = 0$ follows from the fact that $M$ is closed. For every $x, y \in X$, let $m, n \in M$ be such that

$$\|x - m\| < \|\dot{x}\| + \epsilon \quad \text{and} \quad \|y - n\| < \|\dot{y}\| + \epsilon.$$

Then

$$\|\dot{x} + \dot{y}\| \leq \|x + y - m - n\| \leq \|x - m\| + \|x - n\| \leq \|\dot{x}\| + \|\dot{y}\| + 2\epsilon.$$

It follows that $\|\dot{x} + \dot{y}\| \leq \|\dot{x}\| + \|\dot{y}\|$. That proves that (13.7) defines a norm.

The following assertions are not hard to prove in the order in which they appear, and we skip to the last one, which is the only one that may present some difficulty.

Suppose that $\{y_n\}$ is a Cauchy sequence in $X/M$. To show that the sequence converges, it suffices to show that a subsequence converges. Passing to a subsequence we may assume that $\|y_n - y_{n+1}\| < 2^{-n}$. Let $x_1 \in X$ be a representative of $y_1$. By assertion (3) in the statement of the theorem, we can find $x_2$ such that $\pi(x_2) = y_2$ and $\|x_1 - x_2\| < \frac{1}{2}$. Continuing inductively, we find a sequence $\{x_n\}$ such that $\|x_n - x_{n+1}\| < 2^{-n}$ for all $n$, and such that $\pi(x_n) = y_n$. If $X$ is complete, the Cauchy sequence $\{x_n\}$ converges to some $x \in X$. By continuity of $\pi$ we have that $y_n \to \pi(x)$. We conclude that $X/M$ is complete. $\square$

**Exercise 13.3.2.** Complete the proof of the theorem.

**Theorem 13.3.3.** *Let $X$ be a normed space and $M \subseteq X$ a closed subspace. Then*

   *1. $M^*$ is isometrically isomorphic to $X^*/M^\perp$.*

   *2. $(X/M)^*$ is isometrically isomorphic to $M^\perp$.*

*Proof.* Define $T : M^* \to X/M^\perp$ by

$$T(m^*) = x^* + M^\perp,$$

where $x^*$ is any bounded extension of $m^*$ to $X$ (which exists by the Hahn-Banach theorem, of course). The operator $T$ is well-defined, because if $y^*$ is another extension of $m^*$, then $x^* - y^* \in M^\perp$. Knowing that $T$ is well-defined, it is an easy matter to prove linearity. $T$ is also surjective, since any $x^* + M^\perp \in X^*/M^\perp$ is given as $x^* + M^\perp = T\left(x^*|_M\right)$. It remains to prove that $T$ is isometric.

Let $m^* \in M^*$ and let $x^*$ be an extension. Then

$$\|T(m^*)\| = \inf\left\{\|x^* + n^*\| : n^* \in M^\perp\right\}.$$

But each functional $x^* + n^*$ extends $m^*$, so $\|x^* + n^*\| \geq \|m^*\|$, whence $\|T(m^*)\| \geq \|m^*\|$. On the other hand, as $n^*$ ranges over all $n^* \in M^\perp$, $x^* + n^*$ ranges over all extensions of $m^*$. By Hahn-Banach, there is some extension, say $y^* = x^* + n_1^*$, such that $\|y^*\| = \|m^*\|$. Thus the infimum is attained and $\|Tm^*\| = \|m^*\|$. That proves the first assertion.

For the second assertion, consider the dual map of $\pi : X \to X/M$, that is the map $\pi^* : (X/M)^* \to X^*$ defined by

$$\pi^*(y^*) = y^* \circ \pi.$$

It is obvious that $\operatorname{Im}(\pi^*) \subseteq M^\perp$, since $\pi$ vanishes on $M$. We need to show that $\pi^*$ is a surjective isometry onto $M^\perp$.

To see that $\pi^*$ is isometric, we compute

$$\|\pi^*(y^*)\| = \sup\left\{|\langle \pi^*(y^*), x\rangle| : \|x\| < 1\right\} = \sup\left\{|\langle y^*, \pi(x)\rangle| : \|x\| < 1\right\}.$$

By Theorem 13.3.1, $\pi$ maps the open unit ball of $X$ onto the open unit ball of $X/M$, so the right-hand side is equal to

$$\sup\{|\langle y^*, y\rangle| : \|y\| < 1\} = \|y^*\|.$$

It remains to show that the range of $\pi^*$ is equal to $M^\perp$. Let $x^* \in M^\perp$. Then $\ker(x^*) \supseteq M$. It follows that if $x_1 - x_2 \in M$, then $\langle x^*, x_1\rangle = \langle x^*, x_2\rangle$, so we may define a functional $f$ on $X/M$ by

$$f(\dot{x}) = \langle x^*, x\rangle.$$

By definition $x^* = f \circ \pi$. The kernel of $f$ is equal to $\pi(\ker(x^*))$, and is closed by Theorem 13.3.1. By Exercise 13.1.12, $f$ is continuous. Thus, $f \in (X/M)^*$ is such that $x^* = \pi^*(f)$, so the range of $\pi^*$ is $M^\perp$. That completes the proof. $\square$

## 13.4    Additional excercises

**Exercise 13.4.1.** Consider the space $X = \ell_3^1$ (i.e., $X = (\mathbb{R}^3, \|\cdot\|_1)$).

1. Let $F \in X^*$ be given by $F(x, y, z) = ax + by + cz$. Prove that $\|F\| = \max\{|a|, |b|, |c|\}$.

2. Let $Y = \{(x, y, z) : z = 0 = x - 3y\}$ be a linear subspace, and let $f : Y \to \mathbb{R}$ be given by $f(x, y, z) = x$ (for $(x, y, z) \in Y$). Find the general form of $F \in X^*$ such that $F|_Y = f$ and $\|F\| = \|f\|$.

**Exercise 13.4.2.** Give a proof of the Hahn-Banach extension theorem (which one?) that works for separable normed spaces, and does not make use of Zorn's lemma.

**Exercise 13.4.3** (Completion of normed and inner product spaces). The goal of this exercise is to provide an alternative proof to Theorems 2.3.1 and 7.1.13. Let $X$ be a normed space. Let $\widehat{X}$ denote the image of $X$ in $X^{**}$ under the map $x \mapsto \hat{x}$. Prove that $\overline{\widehat{X}}$ is the unique completion of $X$. Prove that if $X$ is an inner product space, then the norm of $\overline{\widehat{X}}$ is induced by an inner product, thus it is a Hilbert space.

**Exercise 13.4.4.** Let $X$ be a real vector space with two norms $\|\cdot\|$ and $\|\cdot\|'$. Let $f$ be a linear functional on $X$, and suppose that for every $x \in X$, either $f(x) \le \|x\|$, or $f(x) \le \|x\|'$. Prove that there is some $t \in [0, 1]$ such that

$$f(x) \le t\|x\| + (1 - t)\|x\|' \quad \text{for all } x \in X.$$

(**Hint:** consider the convex sets $\{(a, b) : a < 0, b < 0\}$ and $\{(\|x\| - f(x), \|x\|' - f(x)) : x \in X\}$ in $\mathbb{R}^2$.)

**Exercise 13.4.5.** Prove that the limit of a weakly convergent sequence in a normed space is unique.

**Exercise 13.4.6.** Prove that a weakly convergent sequence in a Banach space is bounded. What happens if the space is not necessarily complete?

**Exercise 13.4.7.** Prove that *any* two convex sets in a finite dimensional normed space can be separated by a hyperplane[1].

**Exercise 13.4.8.** Let $\ell^\infty(\mathbb{Z})$ be the space of all two-sided bounded sequences of real numbers, and let $U : \ell^\infty(\mathbb{Z}) \to \ell^\infty(\mathbb{Z})$ be the bilateral shift operator, given by

$$U(x_n)_{n \in \mathbb{Z}} = (x_{n-1})_{n \in \mathbb{Z}}.$$

Prove that there exists a bounded linear functional $f \in (\ell^\infty(\mathbb{Z}))^*$ such that

---

[1] I took this exercise from [4, Exercise 1.9], where the student can also find a hint to one possible solution.

1. $f(Ux) = f(x)$ for all $x \in \ell^\infty(\mathbb{Z})$,

2. $\inf_n x_n \le f(x) \le \sup_n x_n$ for all $x \in \ell^\infty(\mathbb{Z})$,

3. $f(\mathbf{1}) = 1$, where $\mathbf{1} = (\ldots, 1, 1, 1, \ldots)$.

Such a linear functional is called a *mean* on $\ell^\infty(\mathbb{Z})$. (**Hint:** use Hahn-Banach to separate the sets $Y = \mathrm{Im}(I - U)$ and $K = \{x \in \ell^\infty(\mathbb{Z}) : \inf_n x_n > 0\}$.)

**Exercise 13.4.9.** Let $X$ be a Banach space and let $N$ be a subspace of $X^*$. True or false: $({}^\perp N)^\perp = \overline{N}$ ?

**Exercise 13.4.10.** Let $T \in B(X, Y)$. True or false: $\ker(T)^\perp = \overline{\mathrm{Im}(T^*)}$ ?

**Exercise 13.4.11.** Let $X$ be a reflexive Banach space, and $M$ a closed subspace of $X$. True or false: $M$ is also a reflexive Banach space?

**Exercise 13.4.12.** True or false: a Banach space $X$ is reflexive if and only if $X^*$ is reflexive?

**Exercise 13.4.13.** Which of the following spaces is reflexive?

1. A finite dimensional normed space.

2. $c_0$.

3. $c$.

4. $C([0, 1])$.

5. $\ell^\infty$.

**Exercise 13.4.14.** Prove that for every $a = (a_n)_0^\infty \in \ell^\infty$, the norm of the coset $a + c_0$ in the quotient space $\ell^\infty / c_0$ is equal to

$$\|a + c_0\| = \limsup_{n \to \infty} |a_n|.$$

# Appendix A

## Metric and topological spaces

In this appendix, we collect some facts about metric and topological spaces that are needed in basic functional analysis, and in particular in this book. For the main theorems we provide full proofs. Straightforward propositions are stated without proof, and the reader is advised (and should be able) to fill in the details. Minor but useful additional facts are stated as exercises.

This short appendix is far from being a comprehensive introduction to the basic theory of metric and topological spaces, many important subjects are missing. For a more thorough treatment suitable for self-study and geared for analysis, I recommend the relevant parts of [33] (see also [11] or [20]).

---

## A.1  Metric spaces

### A.1.1  Basic definitions

**Definition A.1.1.** Let $X$ be a set. A **metric** on $X$ is a nonnegative function $d : X \times X \to [0, \infty)$ such that for all $x, y, z \in X$,

1. $d(x, y) = 0$ if and only if $x = y$,

2. $d(x, y) = d(y, x)$,

3. $d(x, z) \leq d(x, y) + d(y, z)$.

A **metric space** is a pair $(X, d)$ where $X$ is a set and $d$ is a metric on $X$.

It is customary to abuse notation slightly and to refer to the space only as the metric space; e.g., we may say "let $X$ be a metric space..." instead of "let $(X, d)$ be a metric space...". Sometimes, the metric on a metric space $X$ is denoted by $d_X$, for emphasis.

**Example A.1.2.** Let $X = \mathbb{R}^k$ and let $d$ be the Euclidean distance, that is

$$d(x, y) = \|x - y\|_2 = \sqrt{\sum_{i=1}^{k} |x_i - y_i|^2}.$$

Then $(X, d) = (\mathbb{R}^k, \| \cdot \|_2)$ is a metric space.

**Example A.1.3.** Let $X = \mathbb{C}$ with the distance

$$d(z, w) = |z - w|.$$

Then $\mathbb{C}$ with this metric can be identified with $\mathbb{R}^2$ with the Euclidean distance.

In any metric space, the condition $d(x, z) \leq d(x, y) + d(y, z)$ is often referred to as the **triangle inequality**. Drawing three points in the plane equipped with the Euclidean distance immediately clarifies why.

The notion of a metric is supposed to serve as a measure of the distance between points in a space. The above example — the Euclidean space — is a space about which we are accustomed to think geometrically. The power of the theory of metric spaces is that it supplies us with a framework in which to study abstract spaces as geometrical objects.

**Example A.1.4.** Consider the space $C_{\mathbb{R}}([a, b])$ of continuous real-valued functions on the interval $[a, b]$. It is straightforward to check that

$$d(f, g) = \|f - g\|_\infty,$$

defines a metric on $C_{\mathbb{R}}([a, b])$, where

$$\|f - g\|_\infty = \sup_{t \in [a,b]} |f(t) - g(t)|$$

for $f, g \in C_{\mathbb{R}}([a, b])$.

**Example A.1.5.** Let $(X, d)$ be a metric space and let $Y \subseteq X$ be any subset. If we define $d_Y = d\big|_{Y \times Y}$, then $(Y, d_Y)$ is also a metric space.

Whenever $X$ is a metric space and $Y \subseteq X$, we consider $Y$ as a metric space with the above metric, referred to as the metric **induced from** $(X, d)$. In this context, $Y$ is also called a **subspace** of $X$. In particular, if we have a subset $E$ of $\mathbb{R}^k$ or of $\mathbb{C}$ under consideration, then, unless stated otherwise, we will consider $E$ as a metric space with the restriction of the Euclidean distance.

Henceforth $(X, d)$ will denote a metric space.

**Definition A.1.6.** Given $x \in X$ and $r > 0$, the set

$$B_r(x) = \{w \in X : d(x, w) < r\}$$

is called the **open ball** with center $x$ and radius $r$.

**Definition A.1.7.** A subset $U \subseteq X$ is said to be **open** (or an **open set**) if for all $x \in U$, there exists $r > 0$ such that $B_r(x) \subseteq U$.

**Proposition A.1.8.** *The trivial subsets $\emptyset, X$ of $X$ are open. If $U_1$ and $U_2$ are open, then $U_1 \cap U_2$ is open. If $\{U_i\}$ is a family of open sets, then $\cup_i U_i$ is open.*

**Exercise A.1.9.** Show that a nonempty subset of a metric space is open if and only if it is the union of open balls.

**Proposition A.1.10.** *Let $X$ be a metric space and let $Y$ be a subspace of $X$ with the induced metric $d_Y$ as in Example A.1.5 . A set $E \subseteq Y$ is open in $(Y, d_Y)$ if and only if there exists an open set $U$ in $X$ such that $E = U \cap Y$.*

In the theory of metric spaces, just as in real analysis, sequences of points play a central role.

**Definition A.1.11.** Let $\{x_n\}_{n=1}^{\infty}$ be a sequence of points in $X$, and let $x \in X$. The sequence is said to **converge to** $x$, if $\lim_{n \to \infty} d(x_n, x) = 0$. In this case, we say that $x$ is the **limit** of the sequence, and we write

$$\lim_{n \to \infty} x_n = x,$$

or $x_n \to x$.

**Example A.1.12.** Consider $C_{\mathbb{R}}([a, b])$ with the metric given in Example A.1.4. It follows from the definitions that a sequence of functions $\{f_n\}_{n=1}^{\infty} \subset X$ converges to $f$ if and only if the sequence $f_n$ converges uniformly to $f$.

**Exercise A.1.13.** Prove that the limit of a sequence in a metric space is unique, in the sense that if $x_n \to x$ and at the same time $x_n \to y$, then $x = y$.

**Exercise A.1.14.** Prove that a sequence $\{x_n\}_{n=1}^{\infty}$ in a metric space is convergent to $x$ if and only if every subsequence of $\{x_n\}_{n=1}^{\infty}$ has a convergent sub-subsequence which converges to $x$.

**Definition A.1.15.** A subset $F \subseteq X$ is said to be **closed** if whenever $\{x_n\}_{n=1}^{\infty}$ is a sequence of points in $F$ that converges to some point $x \in X$, then the point $x$ is also in $F$.

**Example A.1.16.** The open ball $B_r(x)$ in a metric space may or may not be closed. If $X = \mathbb{R}^k$ with the Euclidean distance, then

$$B_r(0) = \{x \in \mathbb{R}^k : \|x\|_2 < r\}$$

is not closed for any $r > 0$. However, the so-called **closed ball**

$$\{x \in \mathbb{R}^k : \|x\|_2 \leq r\}$$

is closed, as the terminology suggests.

**Proposition A.1.17.** *A subset $F \subseteq X$ is closed if and only if its complement $X \setminus F$ is open.*

**Definition A.1.18.** Let $E$ be a subset of $X$. The **closure** of $E$ is defined to be the set

$$\overline{E} = \bigcap_{\substack{F \supseteq E \\ F \text{ is closed}}} F.$$

The **interior** of $E$ is defined to be the set

$$\text{int}(E) = \bigcup_{\substack{U \subseteq E \\ U \text{ is open}}} U.$$

From the definition it follows that $\overline{E}$ is closed, and is the smallest closed set containing $E$. Likewise, $\text{int}(E)$ is open, and is the largest open set contained in $E$.

**Exercise A.1.19.** Let $E$ be a subset of a metric space. Then, $x \in \overline{E}$ if and only if there is a sequence of points $\{x_n\}_{n=1}^{\infty}$ contained in $E$ such that $x_n \to x$. Also, $x \in \text{int}(E)$ if and only if there exists some $r > 0$ such that $B_r(x) \subseteq E$.

**Exercise A.1.20.** Let $E$ be a subset of a metric space. Then, $x \in \overline{E}$ if and only if, for every $r > 0$, there exists some $y \in E$ which is also in $B_r(x)$. Also, $x \in \text{int}(E)$ if and only if whenever $x_n \to x$, there exists some $N$ such that $x_n \in E$ for all $n \geq N$.

**Definition A.1.21.** If $E \subseteq X$ and $\overline{E} = X$, then $E$ is said to be **dense** in $X$. If $X$ contains a countable dense subset, then $X$ is said to be **separable**.

**Example A.1.22.** The set of rational numbers $\mathbb{Q}$ is a nonclosed subset of $\mathbb{R}$. The closure of $\mathbb{Q}$ is $\mathbb{R}$. In particular, $\mathbb{R}$ is separable.

## A.1.2 Continuous maps on metric spaces

In metric spaces, the notion of continuity is defined precisely as it is for real-valued functions on an interval, with the metric replacing the absolute value of the difference between numbers.

**Definition A.1.23.** Let $(X, d_X)$ and $(Y, d_Y)$ be two metric spaces and let $x \in X$. A function $f : X \to Y$ is said to be **continuous at** $x$ if for every $\epsilon > 0$, there exists $\delta > 0$ such that for all $w \in X$,

$$d_X(x, w) < \delta \quad \text{implies that} \quad d_Y(f(x), f(w)) < \epsilon.$$

The function $f$ is said to be **continuous on** $X$ (or simply **continuous**) if it is continuous at every point of $X$.

The set of all continuous functions from $X$ to $Y$ is denoted by $C(X, Y)$. The space of all complex-valued continuous functions on $X$, that is, $C(X, \mathbb{C})$ is denoted by $C(X)$, and the set of real-valued continuous functions on $X$ by $C_{\mathbb{R}}(X)$.

Recall that given a function $f : X \to Y$ and a set $U \subseteq Y$, the **inverse image** of $U$ under $f$ is the set $f^{-1}(U) = \{x \in X : f(x) \in U\}$.

**Proposition A.1.24.** *If $f : X \to Y$ is a function between metric spaces, then $f$ is continuous on $X$ if and only if for every open $U \subseteq Y$, the inverse image $f^{-1}(U)$ is open in $X$.*

**Proposition A.1.25.** *A function $f$ is continuous at $x \in X$ if and only if for every sequence $\{x_n\}_{n=1}^{\infty} \subseteq X$,*

$$x_n \to x \quad \text{implies that} \quad f(x_n) \to f(x).$$

**Definition A.1.26.** Let $(X, d_X)$ and $(Y, d_Y)$ be two metric spaces. A function $f : X \to Y$ is said to be **uniformly continuous** if for every $\epsilon > 0$, there exists $\delta > 0$ such that for all $x, w \in X$,

$$d_X(x, w) < \delta \quad \text{implies that} \quad d_Y(f(x), f(w)) < \epsilon.$$

---

## A.2 Completeness

### A.2.1 Cauchy sequences and completeness

In what follows, $(X, d)$ continues to denote a metric space.

**Definition A.2.1.** A sequence $\{x_n\}_{n=1}^{\infty} \subseteq X$ is said to be a **Cauchy sequence** if for every $\epsilon > 0$, there exists an integer $N$, such that

$$d(x_m, x_n) < \epsilon \quad \text{for all} \ m, n \geq N.$$

The condition in the definition is sometimes written compactly as

$$d(x_m, x_n) \xrightarrow{m,n \to \infty} 0.$$

**Exercise A.2.2.** Prove that every convergent sequence is a Cauchy sequence.

**Example A.2.3.** A familiar fact from real analysis is that if $\{x_n\}_{n=1}^{\infty}$ is a Cauchy sequence in $\mathbb{R}$, then it is convergent in $\mathbb{R}$. On the other hand, consider the sequence $x_n = \frac{1}{n}$ as a sequence in the metric space $(0, 1)$ with the usual metric $d(x, y) = |x - y|$. Since this sequence converges to 0 in $\mathbb{R}$, it is a Cauchy sequence. However, 0 is not a point in the space $(0, 1)$, so $\{x_n\}_{n=1}^{\infty}$ is a Cauchy sequence that is not convergent in the space $(0, 1)$.

**Exercise A.2.4.** Prove that every Cauchy sequence is bounded.

**Exercise A.2.5.** Prove that a Cauchy sequence which has a convergent subsequence is convergent.

**Definition A.2.6.** A metric space $(X, d)$ is said to be **complete** if every Cauchy sequence in $X$ is convergent.

**Example A.2.7.** The set of real numbers $\mathbb{R}$ is complete. So are all the Euclidean space $\mathbb{R}^k$, and in particular $\mathbb{C}$.

**Example A.2.8.** Consider again the space $C_{\mathbb{R}}([a,b])$ of continuous functions with the distance $d(f,g) = \sup_{t \in [a,b]} |f(t) - g(t)|$. We claim that this space is complete. Indeed, if $\{f_n\}_{n=1}^{\infty}$ is a Cauchy sequence in $C_{\mathbb{R}}([a,b])$, then for every $t \in [a,b]$, the sequence $\{f_n(t)\}_{n=1}^{\infty}$ is a Cauchy sequence in $\mathbb{R}$, and therefore converges. Thus, we can define a function $f : [a,b] \to \mathbb{R}$ by $f(t) = \lim_n f_n(t)$, and it follows that $f_n \to f$ uniformly. Now, it is standard exercise in real analysis to show that the uniform limit of continuous functions is continuous, thus $f \in C_{\mathbb{R}}([a,b])$, and this shows that $C_{\mathbb{R}}([a,b])$ is complete. In pretty much the same way one shows that $C([a,b])$ — the space of continuous complex-valued functions on $[a,b]$ — is complete.

**Exercise A.2.9.** Prove that if $X$ is a complete metric space and $Y \subseteq X$, then $Y$ is complete in the induced metric if and only if $Y$ is closed in $X$.

## A.2.2 The completion of a metric space

**Theorem A.2.10.** *For every metric space $X$, there exists a complete metric space $Y$ and a map $j : X \to Y$ such that*

*1. $d_Y(j(x), j(w)) = d_X(x,w)$ for all $x, w \in X$.*

*2. $j(X)$ is dense in $Y$.*

*If $Z$ is another complete metric space and $k : X \to Z$ is a map satisfying the above two conditions, then there is a bijective map $f : Y \to Z$ such that $d_Z(f(y_1), f(y_2)) = d_Y(y_1, y_2)$ for all $y_1, y_2 \in Y$ and $f(j(x)) = k(x)$ for all $x \in X$.*

**Remarks A.2.11.** Before the proof we make the following remarks.

1. The space $Y$ is said to be **the metric space completion of** $X$, or simply **the completion of** $X$.

2. The last assertion of the theorem says that $Y$ is essentially unique, since $Z$ is a metric space which is in a bijective and metric preserving correspondence with $Y$. We say that $Z$ is **isometric** to $Y$.

3. One usually identifies $X$ with $j(X)$ and then the theorem is stated as follows: *There exists a unique complete metric space $Y$ which contains $X$ as a dense subspace.*

*Proof.* Let $Y_0$ be the set consisting of all Cauchy sequences in $X$. To be precise, let $X^{\mathbb{N}}$ be the set of all sequences in $X$

$$X^{\mathbb{N}} = \{\mathbf{x} = \{x_k\}_{k=1}^{\infty} : x_k \in X \text{ for all } k = 1, 2, \ldots\},$$

and let $Y_0$ be the subset of $X^{\mathbb{N}}$ consisting of all Cauchy sequences.

We define an equivalence relation on $Y_0$ by declaring that two Cauchy sequences $\mathbf{x} = \{x_k\}_{k=1}^{\infty}$ and $\mathbf{w} = \{w_k\}_{k=1}^{\infty}$ are equivalent, denoted $\mathbf{x} \sim \mathbf{w}$, if and only if $\lim_{k \to \infty} d_X(x_k, w_k) = 0$. We define $Y$ to be the set $Y_0/ \sim$ of all equivalence classes. Thus $Y = \{\dot{\mathbf{x}} : \mathbf{x} \in Y_0\}$, where $\dot{\mathbf{x}}$ denotes the equivalence class of $\mathbf{x}$, that is

$$\dot{\mathbf{x}} = \{\mathbf{w} = \{w_k\}_{k=1}^{\infty} \in Y_0 : \mathbf{w} \sim \mathbf{x}\}.$$

We now define a metric on $Y$. For $\mathbf{x}, \mathbf{w} \in Y_0$, we define

$$d_Y(\dot{\mathbf{x}}, \dot{\mathbf{w}}) = \limsup_{k \to \infty} d_X(x_k, w_k). \tag{A.1}$$

If $\mathbf{x} \sim \mathbf{x}'$ and $\mathbf{w} \sim \mathbf{w}'$, then a simple estimation using the triangle inequality shows that $\limsup_{k \to \infty} d_X(x_k, w_k) = \limsup_{k \to \infty} d_X(x_k', w_k')$, so $d_Y$ is well-defined. It is straightforward to check that $d_Y$ is indeed a metric.

We define $j : X \to Y$ by letting $j(x)$ be the equivalence class of the constant sequence $\{x_k\}_{k=1}^{\infty}$ with $x_k = x$ for all $k$. Then, $d_Y(j(x), j(w)) = d_X(x, w)$ for all $x, w \in X$.

Next, we show that $j(X)$ is dense in $Y$. Let $\dot{\mathbf{x}} \in Y$, with representative $\mathbf{x} = \{x_k\}_{k=1}^{\infty}$. Given $\epsilon > 0$, let $k_0$ be such that $d_X(x_k, x_l) < \epsilon$ for all $k, l \geq k_0$. Then $d_Y(j(x_{k_0}), \dot{\mathbf{x}}) \leq \epsilon$. As this is true for every $\dot{\mathbf{x}} \in Y$, and every $\epsilon > 0$, we conclude that the closure of $j(X)$ is equal to $Y$.

It remains to show that $Y$ is complete. Suppose that $\{\dot{\mathbf{x}}^{(n)}\}_{n=1}^{\infty}$ is a Cauchy sequence in $Y$. We will find an element $\dot{\mathbf{y}} \in Y$ such that $\dot{\mathbf{x}}^{(n)} \to \dot{\mathbf{y}}$.

Since $j(X)$ is dense in $Y$, we may choose, for every $n$, a point $y_n \in X$ so that $d_Y(j(y_n), \dot{\mathbf{x}}^{(n)}) < \frac{1}{n}$. Define $\mathbf{y} = \{y_k\}_{k=1}^{\infty} \in X^{\mathbb{N}}$. This is a Cauchy sequence, because

$$d_X(y_m, y_n) = d_Y(j(y_m), j(y_n))$$
$$\leq d_Y\left(j(y_m), \dot{\mathbf{x}}^{(m)}\right) + d_Y\left(\dot{\mathbf{x}}^{(m)}, \dot{\mathbf{x}}^{(n)}\right) + d_Y\left(\dot{\mathbf{x}}^{(n)}, j(y_n)\right)$$
$$< \frac{1}{m} + d_Y\left(\dot{\mathbf{x}}^{(m)}, \dot{\mathbf{x}}^{(n)}\right) + \frac{1}{n} \xrightarrow{m,n \to \infty} 0.$$

Finally, to see that $\dot{\mathbf{x}}^{(n)} \to \dot{\mathbf{y}}$, we compute

$$d_Y\left(\dot{\mathbf{x}}^{(n)}, \dot{\mathbf{y}}\right) = \limsup_{k \to \infty} d_X\left(x_k^{(n)}, y_k\right)$$
$$\leq \limsup_{k \to \infty} d_X\left(x_k^{(n)}, y_n\right) + \limsup_{k \to \infty} d_X(y_n, y_k)$$
$$= d_Y\left(\dot{\mathbf{x}}^{(n)}, j(y_n)\right) + \limsup_{k \to \infty} d_X(y_n, y_k)$$
$$< \frac{1}{n} + \limsup_{k \to \infty} d_X(y_n, y_k) \xrightarrow{n \to \infty} 0.$$

That finishes the construction of the completion of $X$. We leave it as an exercise to prove that the completion is unique. $\qquad\square$

**Exercise A.2.12.** Fill in details that you feel are missing (suggestion: prove that $d_Y$ is well-defined and a metric).

**Exercise A.2.13.** Prove that the lim sup in A.1 is actually a limit.

**Example A.2.14.** Prove that $\mathbb{R}$ is the completion of $\mathbb{Q}$.

**Exercise A.2.15.** Let $X$ be a metric space, and let $Y \subset X$ be a dense subspace. If $Z$ is a complete metric space, and if $f : Y \to Z$ is a uniformly continuous map, then there exists a unique uniformly continuous map $\tilde{f} : X \to Z$ such that $\tilde{f}\big|_Y = f$.

## A.2.3   Baire's category theorem

**Definition A.2.16.** A subset $E \subseteq X$ is said to be ***nowhere dense*** (in $X$) if $\text{int}(\overline{E}) = \emptyset$. If $Y \subseteq X$ can be given as a countable union of nowhere dense sets, that is, if there exists a sequence of nowhere dense sets $E_n$ such that $Y = \cup_{n=1}^{\infty} E_n$, then $Y$ is said to be of the ***first category*** in the sense of Baire; otherwise, it is said to be of the ***second category***.

Note that in the above definition, the question whether a subset of a metric space $X$ is nowhere dense (and hence, whether it is of the first or second category) depends on the ambient space $X$. We will say that $X$ ***is of the second category*** if it is of the second category considered as a subspace of itself. In this book, the terminology of first and second category will not be used. In fact, hardly anyone uses this terminology; it has been mentioned only for the purpose of explaining the origin of the name of the following important theorem, which has deep consequences in functional analysis.

**Theorem A.2.17** (Baire's category theorem). *Let $X$ be a nonempty complete metric space. Then $X$ is of the second category. In other words, if $\{E_n\}_{n=1}^{\infty}$ is a countable family such that $X = \cup_{n=1}^{\infty} E_n$, then $\text{int}(\overline{E_n}) \neq \emptyset$ for some $n$.*

*Proof.* We will show that if $\text{int}(\overline{E_n}) = \emptyset$ for all $n$, then $\cup_{n=1}^{\infty} E_n \neq X$. Consider the sequence of sets $G_n = X \setminus \overline{E_n}$. Then every $G_n$ is an open subset of $X$. Moreover, if $\text{int}(\overline{E_n}) = \emptyset$, then $G_n$ is dense (and in particular not empty). We will construct a Cauchy sequence $\{x_n\}_{n=1}^{\infty}$ that converges to a point $x \in \cap_n G_n$, providing a point $x$ not in $\cup_n E_n \subseteq \cup_n \overline{E_n} = X \setminus (\cap_n G_n)$.

Choose any $x_1 \in G_1$, and let $r_1 > 0$ be such that $\overline{B_{r_1}(x_1)} \subseteq G_1$. We continue choosing a sequence $x_1, x_2, \ldots \in X$ and $r_1, r_2, \ldots > 0$ inductively. Suppose that we have chosen $x_1, \ldots, x_n$ and $r_1, \ldots, r_n$ in such a way that $B_{r_1}(x_1) \supseteq \ldots \supseteq B_{r_n}(x_n)$ and

$$\overline{B_{r_n}(x_n)} \subseteq G_1 \cap \cdots \cap G_n.$$

Since $G_{n+1}$ is open and dense, we choose some $x_{n+1} \in B_{r_n}(x_n) \cap G_{n+1}$, and we pick $r_{n+1} < \frac{r_n}{2}$ such that

$$\overline{B_{r_{n+1}}(x_{n+1})} \subseteq B_{r_n}(x_n) \cap G_{n+1}.$$

By construction we have that $\overline{B_{r_{n+1}}(x_{n+1})} \subseteq G_1 \cap \cdots \cap G_{n+1}$.

The sequence $\{x_n\}_{n=1}^\infty$ is a Cauchy sequence. Indeed, if $\epsilon > 0$ is given, we let $N$ be large enough so that $r_N < \frac{\epsilon}{2}$. Then for all $m, n \geq N$, both $x_m$ and $x_n$ are in $B_{r_N}(x_N)$, thus

$$d(x_m, x_n) < 2r_N < \epsilon.$$

As $X$ is a complete metric space, the sequence has a limit $x = \lim_n x_n$. For every $N$, the sequence $\{x_n\}_{n=N}^\infty$ is contained in $B_{r_N}(x_N)$. It follows that $x \in \overline{B_{r_N}(x_N)} \subseteq \cap_{n=1}^N G_n$. Therefore $x \in \cap_{n=1}^\infty G_n$, and the proof is complete. $\qquad\square$

## A.3    Topological spaces

### A.3.1    Basic definitions

It might seem that continuous functions are precisely the right maps to consider in the category of metric spaces. However, note that continuous maps fail to preserve important metric properties. Needless to say, continuous maps need not preserve the metric; but even softer properties may get lost.

**Example A.3.1.** Consider the metric spaces $X = \left(-\frac{\pi}{2}, \frac{\pi}{2}\right)$ and $Y = \mathbb{R}$, both endowed with the standard metric $d(x, y) = |x - y|$. Note that the function $x \mapsto \tan(x)$ is a bijective continuous map of $X$ onto $Y$ with a continuous inverse. However, $Y$ is a complete metric space, while $X$ is not.

The above example leads to the question: *what is the thing that is preserved by continuous maps?* By Proposition A.1.24, the map $f(x) = \tan(x)$ in the above example has the property that $U$ is open in $X$ if and only if $f(U)$ is open in $Y$. This motivates the following definition.

**Definition A.3.2.** Let $X$ be a set. A family $\mathcal{T}$ consisting of subsets of $X$ is said to be a ***topology*** on $X$ if the following properties are satisfied:

1. $\emptyset, X \in \mathcal{T}$,

2. If $U_1, U_2 \in \mathcal{T}$, then $U_1 \cap U_2 \in \mathcal{T}$,

3. If $\{U_i\}_{i \in I}$ is a family of subsets of $X$ and $U_i \in \mathcal{T}$ for all $i \in I$, then $\cup_{i \in I} U_i \in \mathcal{T}$.

The pair $(X, \mathcal{T})$ is then called a ***topological space***. The elements of $\mathcal{T}$ are called ***open sets***.

It is convenient and common to abuse notation and call $X$ itself the topological space, and simply refer to elements of $\mathcal{T}$ as open sets, without explicitly mentioning $\mathcal{T}$. We will also use this convention when convenient.

**Example A.3.3.** Every metric space gives rise to a topological space, with the topology given by the collection of all open sets in the sense of Definition A.1.7 (see Proposition A.1.8). This topology is called the topology *induced* by the metric.

**Example A.3.4.** The space $\mathbb{R}^k$ can be given the Euclidean metric (Example A.1.2), and this gives rise to a family of open sets and hence a topology on $\mathbb{R}^k$. This topology is called the **standard topology on $\mathbb{R}^k$**. There are many other natural metrics one can define on $\mathbb{R}^k$, for example

$$d(x,y) = \|x - y\|_\infty = \max\{|x_i - y_i| : i = 1, \ldots, k\}.$$

If $k > 1$, then this metric is different from the Euclidean distance. However, one may check that both these metrics give rise to the same topology.

Basic propositions in metric spaces become definitions in topological spaces.

**Definition A.3.5.** Let $(X, \mathcal{T})$ be a topological space. A subset $F \subseteq X$ is said to be **closed** if its complement is open, that is, if $X \setminus F \in \mathcal{T}$.

**Definition A.3.6.** Let $(X, \mathcal{T}_X)$ be a topological space and let $Y \subseteq X$. We define a topology on $Y$ by

$$\mathcal{T}_Y = \{U \cap Y : U \in \mathcal{T}_X\}.$$

The topology $\mathcal{T}_Y$ is called the **topology on $Y$ induced by $\mathcal{T}_X$** (or simply the **induced topology**).

Note that this definition is in accordance with Proposition A.1.10. The topological space $Y$ equipped with the induced topology is called a **subspace** of $X$, and the induced topology is also sometimes referred to as the **subspace topology**.

**Definition A.3.7.** Let $(X, \mathcal{T}_X)$ and $(Y, \mathcal{T}_Y)$ be two topological spaces. A function $f : X \to Y$ is said to be **continuous** if $f^{-1}(U) \in \mathcal{T}_X$ for all $U \in \mathcal{T}_Y$. The set of all continuous functions from $X$ to $Y$ is denoted by $C(X, Y)$. One also writes $C_\mathbb{R}(X) = C(X, \mathbb{R})$ and $C(X) = C(X, \mathbb{C})$.

**Definition A.3.8.** If $f : X \to Y$ is a continuous and bijective function, and if the inverse function $f^{-1} : Y \to X$ is also continuous, then $f$ is said to be a **homeomorphism**.

Thus, a bijective function $f$ is a homeomorphism if it satisfies that $f(U)$ is open if and only if $U$ is open. Homeomorphisms are the "isomorphisms" in the world of topological spaces, and topology may be considered as the study of those properties invariant under continuous maps.

**Exercise A.3.9.** If $f : X \to Y$ is a continuous function between topological spaces, and $Z$ is a subspace of $X$, then $f\big|_Z$ is continuous.

**Definition A.3.10.** Let $(X, \mathcal{T}_X)$ and $(Y, \mathcal{T}_Y)$ be two topological spaces. Let

$$X \times Y = \{(x, y) : X \in X, y \in Y\}$$

be the Cartesian product in the usual set theoretic sense. The ***product topology*** on $X \times Y$, denoted by $\mathcal{T}_{X \times Y}$, is the collection of all unions of products $U \times V$, where $U$ is open in $X$ and $V$ is open in $Y$, that is

$$\mathcal{T}_{X \times Y} = \left\{ \bigcup_{i \in I} U_i \times V_i : \{(U_i, V_i)\}_{i \in I} \text{ is a subset of } \mathcal{T}_X \times \mathcal{T}_Y \right\}.$$

**Exercise A.3.11.** Prove that the product topology is indeed a topology on $X \times Y$.

**Exercise A.3.12.** Let $(X, d_X)$ and $(Y, d_Y)$ be two metric spaces with the induced topology. Prove that the function

$$d : (X \times Y) \times (X \times Y) \longrightarrow [0, \infty)$$

given by

$$d\left((x, y), (x', y')\right) = d_X(x, x') + d_Y(y, y')$$

is a metric on $X \times Y$. Prove that this metric induces the product topology on $X \times Y$.

The notions of closure and interior are really topological definitions. We repeat the definitions for definiteness.

**Definition A.3.13.** Let $E$ be a subset of a topological space $X$. The ***closure*** of $E$ is defined to be the set

$$\overline{E} = \bigcap_{\substack{F \supseteq E \\ F \text{ is closed}}} F.$$

The ***interior*** of $E$ is defined to be the set

$$\text{int}(E) = \bigcup_{\substack{U \subseteq E \\ U \text{ is open}}} U.$$

**Definition A.3.14.** Let $E$ be a subset of a topological space $X$. A set $U \subseteq X$ is said to be a ***neighborhood of*** $E$ if $U$ contains an open set $V$ that contains $E$. If $E = \{x\}$, then $U$ is said to be a ***neighborhood of*** $x$.

The reader is warned that oftentimes the word "neighborhood" is used to mean what according to the above definition should be called an "open neighborhood".

**Definition A.3.15.** A topological space $X$ is said to be a ***Hausdorff*** space if for every two distinct points $x, y \in X$, there exist two open sets $U, V \subset X$ such that $x \in U$, $y \in V$ and $U \cap V = \emptyset$.

Thus, a topological space is Hausdorff if and only if every two distinct points have disjoint neighborhoods.

**Exercise A.3.16.** Every metric space is a Hausdorff space.

Using the above result, one can show that there are topologies that are not induced by any metric.

**Example A.3.17.** Consider a set $X = [0,1] \cup \{o\}$, where $o$ is some element not in $[0,1]$. We define a topology on $X$ as follows: $U \subseteq X$ is open if and only if either $U$ is an open subset of $[0,1]$, or $U = \{o\} \cup V$ where $V$ is an open subset of $[0,1]$ containing $(0,r)$ for some $r > 0$. One checks that this is indeed a topology on $X$. Every open set containing 0 intersects every open set that contains $o$, so $X$ is not Hausdorff.

There is a notion of convergence of sequences in topological spaces, the definition is included below for completeness. Sequences in topological spaces are not as useful as they are in metric spaces. There is a more sophisticated notion of *nets* that replaces sequences. The interested reader is referred to [11].

**Definition A.3.18.** Let $\{x_n\}_{n=1}^{\infty}$ be a sequence of points in a topological space $X$, and let $x \in X$. The sequence is said to **converge to** $x$, if for every open set $U$ containing $x$, there exists some $N$ such that

$$x_n \in U \quad \text{for all} \quad n \geq N.$$

In this case, we say that $x$ is a **limit** of the sequence $\{x_n\}_{n=1}^{\infty}$.

**Exercise A.3.19.** Prove that, in a metric space, a sequence converges in the sense of Definition A.1.11 if and only if it converges in the sense of A.3.18.

**Example A.3.20.** Let $X$ be as in Example A.3.17. The sequence $x_n = \frac{1}{n}$ converges to 0, but it also converges to $o$. However, $0 \neq o$. We see that a sequence in a topological space might converge to two different limits.

**Exercise A.3.21.** Prove that in a Hausdorff topological space, the limit of a convergent sequence is unique.

## A.3.2  Compactness

**Definition A.3.22.** A topological space $X$ is said to be **compact** if the following condition holds: whenever there is a family $\{U_i\}_{i \in I}$ of open sets in $X$ such that $X = \cup_{i \in I} U_i$, there exists a finite set of indices $i_1, \ldots, i_n \in I$ such that $X = \cup_{k=1}^{n} U_{i_k}$.

In other words, $X$ is compact if every open cover contains a finite subcover. Here, an **open cover** is a family of open sets $\{U_i\}_{i \in I}$ such that $X = \cup_{i \in I} U_i$.

**Exercise A.3.23.** Let $X$ be a topological space, and let $F \subseteq X$ be a subspace. Prove that if we equip $F$ with the subspace topology, then it is compact if and only if the following condition holds: whenever there is a family $\{U_i\}_{i \in I}$ of open sets in $X$ such that $F \subseteq \cup_{i \in I} U_i$, there exists a finite set of indices $i_1, \ldots, i_n \in I$ such that $F \subseteq \cup_{k=1}^n U_{i_k}$.

If the condition in the above exercise holds, then we refer to $F$ as a *compact* subset of $X$.

**Proposition A.3.24.** *Let $f : X \to Y$ be a continuous function between topological spaces. If $K \subseteq X$ is compact, then $f(K)$ is compact.*

The above proposition, together with Theorem A.3.27 below, implies the following corollary.

**Corollary A.3.25.** *Let $X$ be a nonempty compact topological space. Every continuous function $f : X \to \mathbb{R}$ is bounded, and attains its maximum and minimum.*

**Exercise A.3.26.** Let $X$ be a compact topological space. Then, the space $C(X)$ endowed with
$$d(f,g) = \sup_{x \in X} |f(x) - g(x)|$$
is a complete metric space.

## A.3.3 Compactness in metric spaces

**Theorem A.3.27.** *Let $X$ be a metric space. The following are equivalent:*

1. *$X$ is compact.*

2. *$X$ is complete, and for every $\epsilon > 0$ there exists a finite set of points $x_1, \ldots, x_n \in X$ such that $X = \cup_{k=1}^n B_\epsilon(x_k)$.*

3. *Every sequence $\{x_n\}_{n=1}^\infty \subseteq X$ has a convergent subsequence.*

Before the proof, we remark that a space that satisfies the second part of Condition 2 is said to be ***totally bounded***. A space satisfying Condition 3 is said to be ***sequentially compact***. Thus, the theorem can be rephrased as follows: *a metric space is compact if and only if it is complete and totally bounded, and this happens if and only if it is sequentially compact.*

*Proof.* [1 $\Rightarrow$ 3]: Let $\{x_n\}_{n=1}^\infty$ be a sequence in the compact space $X$, and suppose for the sake of obtaining a contradiction, that it has no convergent subsequence. (This means, in particular, that $\{x_n\}_{n=1}^\infty$ contains infinitely many distinct points.) But then for every $x \in X$, there exists some $r_x > 0$, such that the intersection of $B_{r_x}(x)$ with the sequence contains at most one point, namely: $B_{r_x}(x) \cap \{x_n\}_{n=1}^\infty = \{x\}$ if $x = x_n$ for some $n$, and otherwise

$B_{r_x}(x) \cap \{x_n\}_{n=1}^{\infty} = \emptyset$. But the cover $\{B_{r_x}(x)\}_{x \in X}$ can then have no finite subcover, in contradiction to compactness.

[1 $\Rightarrow$ 2]: Let $\epsilon > 0$ be given. Then we cover $X$ by the family of all balls $B_{\epsilon}(x)$ for all $x \in X$. By compactness, there is a finite subcover, that is, a finite set of points $x_1, \dots, x_n \in X$ such that $X = \cup_{k=1}^{n} B_{\epsilon}(x_k)$. Now, let $\{x_n\}_{n=1}^{\infty} \subseteq X$ be a Cauchy sequence. To prove that it converges, it suffices to prove that it contains a convergent subsequence (Exercise A.2.5). But by the implication [1 $\Rightarrow$ 3], we are done.

[2 $\Rightarrow$ 3]: Assume that Condition 2 holds, and let $\{x_n\}_{n=1}^{\infty}$ be a sequence in $X$. Since $X$ is complete, it suffices to prove that $\{x_n\}_{n=1}^{\infty}$ has a Cauchy subsequence. We may treat the case where $\{x_n\}_{n=1}^{\infty}$ contains infinitely many distinct points, the other case being trivial. By Condition 2 applied to $\epsilon = 1$, we find finitely many balls of radius 1 covering $X$, and therefore at least one of these balls, say, $B_1(y_1)$, contains infinitely many points from the sequence. Let $\{x_{1n}\}_{n=1}^{\infty}$ be the subsequence of $\{x_n\}_{n=1}^{\infty}$ consisting of all elements of the sequence contained in $B_1(y_1)$. Applying the same reasoning with $\epsilon = \frac{1}{2}$, we extract a subsequence $\{x_{2n}\}_{n=1}^{\infty}$ consisting of elements from $\{x_{1n}\}_{n=1}^{\infty}$ contained in some ball $B_{\frac{1}{2}}(y_2)$. Continuing inductively, we find a subsequence $\{x_{kn}\}_{n=1}^{\infty}$ that is contained in a ball with radius $\frac{1}{k}$. Finally, the sequence $\{x_{kk}\}_{k=1}^{\infty}$ is, by construction, a Cauchy subsequence of $\{x_n\}_{n=1}^{\infty}$.

[3 $\Rightarrow$ 2] is the easiest implication and is left to the reader.

Finally, we prove that 2 and 3 together imply 1. Let $\{U_i\}_{i \in I}$ be an open cover of $X$. We claim:

**Claim.** *There exists $r > 0$ such that for every $x \in X$, there exists some $i \in I$ such that $B_r(x) \subseteq U_i$.*

Assuming the claim for the moment, we let $\epsilon = r$ in Condition 2, and find $x_1, \dots, x_n \in X$ such that $X = \cup_{k=1}^{n} B_r(x_k)$. Then, for every $k = 1, \dots, n$, there exists some $i_k \in I$, such that $B_r(x_k) \subseteq U_{i_k}$. But then $X = \cup_{k=1}^{n} B_r(x_k) \subseteq \cup_{k=1}^{n} U_{i_k}$, establishing the existence of a finite subcover, as required.

To complete the proof, we prove the claim. If we assume that there is no such $r > 0$, then we see that there must be a sequence of points $x_n \in X$, and a sequence of radii $r_n \searrow 0$, such that $B_{r_n}(x_n)$ is not contained in any $U_i$. By passing to a subsequence, we may, thanks to Condition 3, assume that there is some $x \in X$ such that $x_n \to x$. Now, there exists some $i_0 \in I$ such that $x \in U_{i_0}$. Since $U_{i_0}$ is open, there is some $\rho > 0$ such that $B_{\rho}(x) \subseteq U_{i_0}$. If $n$ is large enough that $d(x_n, x) < \frac{\rho}{2}$ and also $r_n < \frac{\rho}{2}$, then we conclude that $B_{r_n}(x_n) \subseteq U_{i_0}$, in contradiction to what we assumed. Therefore, there must exist an $r > 0$ such that for all $x$ there exists $i$ for which $B_r(x) \subseteq U_i$.  $\square$

It is now easy to prove the following theorem, which gives a straightforward characterization of compact sets in $\mathbb{R}^k$.

**Theorem A.3.28** (Heine-Borel theorem). *Let $\mathbb{R}^k$ be given the standard topology. A subset $F \subset \mathbb{R}^k$ is compact if and only if $F$ is bounded and closed.*

**Proposition A.3.29.** *Every continuous function from a compact metric space into another metric space is uniformly continuous.*

## A.4   The Arzelà-Ascoli theorem

**Definition A.4.1.** Let $X$ be a topological space. A subset $E \subseteq X$ is said to be **precompact** if its closure $\overline{E}$ is compact.

**Proposition A.4.2.** *A subset $E$ of a metric space $X$ is precompact if and only if every sequence in $E$ has a convergent subsequence.*

To emphasize, the subsequence in the above proposition converges to an element of $X$, which may or may not be in $E$.

The Arzelà-Ascoli theorem gives a characterization of when a subset of $C(X)$ is precompact, for $X$ a compact metric space. We need this theorem in Chapter 9, in order to show that an integral operator on $C([a,b])$ with a continuous kernel is a compact operator.

**Definition A.4.3.** Let $X$ be a metric space. A subset $\mathcal{F} \subseteq C(X)$ is said to be **uniformly bounded** if there exists $M > 0$ such that

$$|f(x)| \leq M \quad \text{for all } x \in X, f \in \mathcal{F}. \tag{A.2}$$

**Definition A.4.4.** Let $X$ be a metric space. A subset $\mathcal{F} \subseteq C(X)$ is said to be **equicontinuous** if for every $\epsilon > 0$, there exists $\delta > 0$, such that for all $f \in \mathcal{F}$ and all $x, y \in X$,

$$d(x,y) < \delta \quad \text{implies that} \quad |f(x) - f(y)| < \epsilon. \tag{A.3}$$

**Exercise A.4.5.** Let $X$ be a metric space. If $\mathcal{F} \subseteq C(X)$ is uniformly bounded and equicontinuous, then $\overline{\mathcal{F}}$ is also uniformly bounded and equicontinuous.

**Theorem A.4.6** (The Arzelà-Ascoli theorem). *Let $X$ be a compact metric space. A set of functions $\mathcal{F} \subseteq C(X)$ is precompact if and only if it is uniformly bounded and equicontinuous.*

In fact, in Chapter 9 we require this result only for the case where $X$ is a closed interval $[a,b]$, and the reader may put $[a,b]$ in place of $X$ if this makes the proof easier to digest.

*Proof.* We will prove that the condition is sufficient for precompactness — this is the implication that we use in Chapter 9. The necessity of this condition is easier, and is left to the reader. (**Hint:** use Condition 2 of Theorem A.3.27 and Proposition A.3.29.)

Assume that $\mathcal{F}$ is uniformly bounded and equicontinuous. By Exercise

A.4.5, $\overline{\mathcal{F}}$ is also uniformly bounded and equicontinuous, thus we may as well assume that $\mathcal{F}$ is closed and our goal is to prove that $\mathcal{F}$ is compact.

By Theorem A.3.27, to show that $\mathcal{F}$ is compact it is enough to show that it is complete and totally bounded. Since $C(X)$ is complete, so is every closed subset of it. It remains to prove — and this is the main part of the proof — that $\mathcal{F}$ is totally bounded.

Fixing $r > 0$, we will find finitely many balls of radius $r$ that cover $\mathcal{F}$. Let $M$ be the constant given by (A.2). Choose complex numbers $z_1, \ldots, z_m \in \{z \in \mathbb{C} : |z| \leq M\}$ such that

$$\{z \in \mathbb{C} : |z| \leq M\} \subseteq \bigcup_{i=1}^{m} \left\{ z \in \mathbb{C} : |z - z_i| < \frac{r}{4} \right\}.$$

Let $\delta$ be the positive number corresponding to $\epsilon = \frac{r}{4}$ in (A.3), and let $x_1, \ldots, x_n \in X$ be points such that

$$X \subseteq \cup_{j=1}^{n} B_\delta(x_j).$$

For every choice of $i_1, \ldots, i_n \in \{1, \ldots, m\}$, put

$$A_{i_1, \ldots, i_n} = \left\{ f \in \mathcal{F} : |f(x_j) - z_{i_j}| < \frac{r}{4} \text{ for all } j = 1, \ldots, n \right\}.$$

The sets $A_{i_1, \ldots, i_n}$ cover $\mathcal{F}$:

$$\mathcal{F} = \bigcup_{i_1, \ldots, i_n = 1}^{m} A_{i_1, \ldots, i_n}.$$

It remains to show that every $A_{i_1, \ldots, i_n}$ is contained in a ball of radius $r$. Indeed, if $A_{i_1, \ldots, i_n} \neq \emptyset$, choose any $f \in A_{i_1, \ldots, i_n}$. To show that $A_{i_1, \ldots, i_n} \subseteq B_r(f)$, we take $g \in A_{i_1, \ldots, i_n}$ and show that $\|f - g\|_\infty < r$. To this end, let $x \in X$ and find $j \in \{1, \ldots, n\}$ such that $x \in B_\delta(x_j)$. Then

$$
\begin{aligned}
&|f(x) - g(x)| \\
&\leq |f(x) - f(x_j)| + |f(x_j) - g(x_j)| + |g(x_j) - g(x)| \\
&< \frac{r}{4} + \frac{2r}{4} + \frac{r}{4} = r.
\end{aligned}
$$

That completes the proof. $\qquad\square$

# Bibliography

[1] J. Agler. Nevanlinna-Pick interpolation on Sobolev space. *Proc. Amer. Math. Soc.*, 108:341–351, 1990.

[2] J. Agler and J.E. M$^c$Carthy. *Pick Interpolation and Hilbert Function Spaces*. American Mathematical Society, 2002.

[3] W.B. Arveson. *A Short Course on Spectral Theory*. Springer, 2002.

[4] H. Brezis. *Functional Analysis, Sobolev Spaces and Partial Differential Equations*. Springer, 2011.

[5] L. Carleson. On convergence and growth of partial sums of Fourier series. *Acta Math.*, 116:135–157, 1966.

[6] C.L. Devito. *Functional Analysis and Linear Operator Theory*. Addison Wesley Pub Co., 1990.

[7] J. Dieudonne. *History of Functional Analysis*. North-Holland, 1983.

[8] R.G. Douglas. *Banach Algebra Techniques in Operator Theory*. Springer, 2013.

[9] Y. Eidelman, V. Milman, and A. Tsolomitis. *Functional Analysis: An Introduction*. American Mathematical Society, 2004.

[10] L.C. Evans. *Partial Differential Equations*. American Mathematical Society, 2010.

[11] G.B. Folland. *Real Analysis: Modern Techniques and their Applications*. Wiley, 1999.

[12] I. Gohberg and S. Goldberg. *Basic Operator Theory*. Birkhäuser, 2001.

[13] M. Hasse. *Functional Analysis: An Elementary Introduction*. American Mathematical Society, 2014.

[14] J.C. Holladay. A note on the Stone-Weierstrass theorem for quaternions. *Proc. Amer. Math. Soc.*, 8:656–657, 1957.

[15] P. Jordan and J. von Neumann. On inner products in linear, metric spaces. *Ann. of Math.*, 36:719–723, 1935.

[16]  R.V. Kadison and J.R. Ringrose. *Fundamentals of the Theory of Operator Algebras, Vol. I.* American Mathematical Society, 1997.

[17]  M. Kline. *Mathematical Thought from Ancient to Modern Times.* Oxford University Press, 1972.

[18]  P. Lax. *Functional Analysis.* Wiley Interscience, 2002.

[19]  B.D. MacCluer. *Elementary Functional Analysis.* Springer, 2008.

[20]  J. Munkres. *Topology.* Pearson, 2000.

[21]  B. Paneah. On a problem in integral geometry connected to the Dirichlet problem for hyperbolic equations. *Internat. Math. Res. Notices*, (5):213–222, 1997.

[22]  B. Paneah. Dynamical approach to some problems in integral geometry. *Trans. Amer. Math. Soc.*, 356:2757–2780, 2004.

[23]  B. Paneah. On the general theory of the Cauchy type functional equations with applications in analysis. *Aequationes Math.*, 74:119–157, 2007.

[24]  M. Reed and B. Simon. *Functional Analysis (Methods of Modern Mathematical Physics, Vol. I).* Academic Press, 1980.

[25]  D. Reem. Character analysis using Fourier series. *Amer. Math. Monthly*, 119:245–246, 2012.

[26]  F. Riesz and B. Sz.-Nagy. *Functional Analysis.* Dover Publications, 1990.

[27]  W. Rudin. *Functional Analysis.* McGraw-Hill, 1973.

[28]  W. Rudin. *Principles of Mathematical Analysis.* McGraw-Hill, 1976.

[29]  W. Rudin. *Real and Complex Analysis.* McGraw-Hill, 1986.

[30]  D. Sarason. *Complex Function Theory.* American Mathematical Society, 2007.

[31]  O.M. Shalit. Noncommutative analysis (blog). `https://noncommutativeanalysis.wordpress.com`.

[32]  O.M. Shalit. Conjugacy of P-configurations and nonlinear solutions to a certain conditional Cauchy equation. *Banach J. Math. Anal.*, 3:28—-35, 2009.

[33]  G.F. Simmons. *Introduction to Topology and Modern Analysis.* Krieger Publishing Company, 2003.

[34]  F.G. Tricomi. *Integral Equations.* Dover Publications, 1985.

[35]  A. Zygmund. *Trigonometric Series.* Cambridge University Press, 1985.

# Index

Printed in the United States
by Baker & Taylor Publisher Services